THE MEAT BUSINESS
Devouring a Hungry Planet

Edited by
Geoff Tansey and Joyce D'Silva

Earthscan Publications Ltd, London

First published in the UK in 1999 by
Earthscan Publications Ltd

A catalogue record for this book is available from the British Library

ISBN: 1 85383 603 6 paperback
1 85383 623 0 hardback

Typesetting by PCS Mapping & DTP, Newcastle upon Tyne
Printed and bound by Creative Print and Design (Wales), Ebbw Vale.
Cover design by Declan Buckley

For a full list of publications please contact:
Earthscan Publications Ltd
120 Pentonville Road
London, N1 9JN, UK
Tel: +44 (0)171 278 0433
Fax: +44 (0)171 278 1142
Email: earthinfo@earthscan.co.uk
http://www.earthscan.co.uk

Earthscan is an editorially independent subsidiary of Kogan Page Ltd and publishes in association with WWF-UK and the International Institute for Environment and Development

This book is printed on elemental chlorine free paper

Contents

PART IV HUMAN WELL-BEING

PART V FARMING THREATS AND OPPORTUNITIES

PART VI TRADE, WELFARE AND VALUES

Acronyms and Abbreviations

AICR	American Institute for Cancer Research
APEDA	Agriculture and Processed Food Export Development Authority
AWBI	Animal Welfare Board of India
BSE	bovine spongiform encephalopathy
BST	bovine somatotrophin
CARTMAN	Centre for Action Research and Technology and Technology for Man, Animal and Nature
CBD	Convention on Biological Diversity
CEEC	Central and Eastern European Countries
CHD	coronary heart disease
CITES	Convention on International Trade in Endangered Species of Wild Flora and Fauna
CIWF	Compassion in World Farming
COMA	Committee on Medical Aspects of Food Policy
nvCJD	new variant Creutzfeldt-Jakob disease
CoE	Council of Europe
ECU	European Currency Unit (replaced by the Euro from Jan 1999)
EU	European Union
FoE	Friends of the Earth
GATT	General Agreement on Tariffs and Trade
GE	genetically engineered
GNP	Gross National Product
HSUS	Humane Society of the United States
IARC	International Agency for Research on Cancer
IFOAM	International Federation of Organic Agricultural Movements
IMF	International Monetary Fund
MAFF	Ministry of Agriculture, Fisheries and Food
MAI	Multilateral Agreement on Investment
MEA	Multilateral Environment Agreements

MEP	Member of European Parliament
MP	Member of Parliament
NAFTA	North America Free Trade Agreement
NAS	National Academy of Sciences
NFU	National Farmers Union
NHS	National Health Service
NGO	non-governmental organization
OECD	Organisation for Economic Co-operation and Development
RSPCA	Royal Society for the Prevention of Cruelty to Animals
SAFE	Sustainable Agriculture, Food and Environment
SAP	Structural Adjustment Programme
SPS	Sanitary and Phytosanitary Measures
TBT	Technical Barriers to Trade
TED	Turtle Exclusion Device
TRIPs	Trade Related Aspects of Intellectual Property Rights
UNICEF	United Nations Children's Fund
USDA	United States Department of Agriculture
WCRF	World Cancer Research Fund
WHO	World Health Organisation
WSPA	World Society for the Protection of Animals
WTO	World Trade Organization
WWF	World Wide Fund For Nature

About the Contributors

Dennis Avery is Director of the Center for Global Food Issues at the Hudson Institute, USA. After graduating in agricultural economics he became the senior agricultural analyst in the US Department of State. He wrote *Saving the Planet with Pesticides and Plastics, the Environmental Triumph of High Yield Agriculture* (Hudson Institute, 1995) and has a weekly syndicated newspaper column.

Geoffrey Cannon is Director of Science and Health Policy, World Cancer Research Fund (WCRF), the one UK charity devoted to the prevention of cancer by means of healthy diets. In 1988 Geoffrey established the Caroline Walker Trust, which aims to improve public health by means of good food, and he is currently Chairman of the National Food Alliance. He has written several books, including *Superbug* (Virgin, 1995) on antibiotics.

Barry Coates became Director of the World Development Movement (WDM) in 1996. Before this he was head of Development Policy at the World Wide Fund For Nature (WWF-UK), where he chaired an international group on sustainable development and represented WWF at the Earth Summit and Commission on Sustainable Development. Barry has a Masters in Public and Private Management from Yale University, has advised governments on public policy and promoted economic development for the government of Western Samoa.

Janice Cox worked for seven years for the World Society for the Protection of Animals as Regional Director for Europe and subsequently for Central and Eastern Europe and as International Legislative Adviser. She now runs, AnimalKind, a humane education organisation. She is also working to set up the World Animal Net, which will help animal organizations around the world network keep in touch with each other, share experiences and collectively work for the common good of animals.

Joyce D'Silva grew up on a farm in Ireland. She graduated from Trinity College, Dublin, taught in India and later in the UK. She has worked for Compassion in World Farming (CIWF) since 1985 and became Director in 1992. She edits the CIWF quarterly magazine, *Agscene*, and has authored chapters on genetic engineering for two books, including the recently published *Animal Biotechnology and Ethics* (Chapman & Hall, 1998).

Michael Fox is the senior scholar, bioethics, of the Humane Society of the United States (HSUS). A vet by training, he writes a syndicated newspaper column called 'Ask Your Animal Doctor' and has written over 40 books, including *Agricide: the Hidden Farm and Food Crisis that Affects Us All* (reprinted 1996) and *Eating with Conscience: The Bioethics of Food* (1997)

Maneka Gandhi became Minister for Welfare in the Indian government in March 1998. She has already introduced restrictions on imports of laboratory animals for research. During a previous stint as Minister of State for the Environment she negotiated the Montreal Protocol for the Indian government and created authorities to control zoos and vivisection in India. She is heavily involved in animal welfare work and is Founder and Chairperson of People for Animals, which runs animal shelters throughout India. She has published several books, including *The Animal Laws of India*, writes a syndicated animal welfare column for 15 national newspapers and produces and writes two weekly programmes on Indian TV. She is a recipient of the Royal Society for the Prevention of Cruelty to Animals' (RSPCA) Lord Erskine Award (1992) and the Marchig Prize for Animal Welfare (1997).

Mark Gold worked for CIWF from 1978–83 and became National Organiser. He was Director of Animal Aid for 12 years and is now a part-time Special Projects Coordinator there. He has written four books on animal issues: *Assault and Battery: What Factory Farming Means for Humans and Animals* (Pluto Press, 1983), *Living Without Cruelty* (Green Point, 1988), *Animal Rights: Extending the Circle of Compassion* (John Carpenter, 1995) and *Animal Century*, (John Carpenter, 1998).

Patrick Holden has farmed organically for 25 years. He was a founder member and first Chair of British Organic Farmers. Since 1987, he has been a member of the UK Government Organic Standards Committee (UKROFS). He is a member of the Agriculture Reform Group and became Director of the Soil Association in 1995.

Tim Lang is Professor of Food Policy and Director of the Centre for Food Policy at Thames Valley University. He was Director of Parents for Safe Food (1990–94) and of the London Food Commission (1984–90), a former Chair of the Sustainable Agriculture, Food and Environment (SAFE) Alliance and is a Trustee of the National Food Alliance and of Friends of the Earth (FoE). He has co-authored seven books, including, with Colin Hines, *The New Protectionism* (Earthscan, 1993) and with Yiannis Gabriel, *The Unmanageable Consumer* (Sage, 1995).

José Lutzenberger is an agronomist who worked for BASF in Germany, in Northern and South America, the Caribbean and in North Africa for a total of 13 years. In 1970 he left to start a campaign against the overuse of agrochemicals and became one of the most formidable fighters against the overuse of agrochemicals in the world and is called the father of the environmental movement in Brazil. In 1988 he received the Right Livelihood Award (alternative Nobel Prize) in Stockholm and from 1990–92 he was National Secretary for the Environment to the President of Brazil.

Philip Lymbery has worked in animal welfare since 1983 and as a Campaigns Director for CIWF since 1993. He authored the UK's first report on the welfare of farmed fish and more recently *Beyond the Battery Cage* (CIWF, 1997), a positive appraisal of the welfare needs of hens. He is also a member of the Soil Association Organic Livestock Standards Committee.

Atherton Martin is currently Executive Director of the Development Institute, a research, consulting and advocacy organization focusing on sustainable development. He was Minister of Agriculture of Dominica in 1979 and General Secretary of the Dominica Farmers' Union from 1978–83. He is also President of the Dominica Conservation Association and the Dominica Hotel and Tourism Association. He has written several publications on the effects of trade liberalization on small island states and the sustainable development options for such states.

Chris Mullin is an author and journalist and has been MP (Labour) for Sunderland South since 1987. He is Chairman of the Home Affairs Select Committee. He has introduced three Ten Minute Rule Bills on farm animal welfare issues: to ban the export of calves to veal crates, to review factory farming and to improve the welfare of pigs.

Hugh Raven is Convenor of the UK Government's Green Globe Task Force, which advises the Foreign Secretary, the Environment Minister and the Secretary of State for International Development on International Environment Policy. He is Chair of SERA, the Labour Party's Environmental Affiliate and a trustee of the Soil Association, the UK's leading organic food and farming organization, and the Royal Society for the Protection of Birds (RSPB).

Julie Sheppard is the Senior Public Affairs Officer at the Consumers Association.

Vandana Shiva trained as a physicist and her PhD was in quantum theory. In 1982 she founded the Research Foundation for Science, Technology and Ecology, dedicated to independent research on ecological and social issues in partnership with local communities. In 1991 she founded Navdanya, a national movement to protect the diversity and integrity of living resources in India. Her books include *Ecofeminism* (Zed Books, 1993) and *Biopiracy* (Green Books, 1999). Among her awards are the Global 500 Roll of Honour 1993 and the Right Livelihood Award 1993.

Peter Stevenson is the Political and Legal Director of CIWF. He has drafted many Private Member's Bills, two of which – the ban on sow stalls and tethers and the Welfare of Animals at Slaughter Act 1991 – have become law. Peter wrote *A Far Cry From Noah* (Green Print, 1994), the only book written on the export of live farm animals, along with a number of highly acclaimed reports on farm animal welfare. Peter also specializes in the ethical and legal issues surrounding the patenting of genetically engineered animals.

Geoff Tansey co-wrote the award winning book *The Food System – a guide* (Earthscan, 1995). He is a full-time freelance writer, broadcaster and consultant specializing in food and farming issues. Since 1996, he has been honorary visiting professor in food policy at Leeds Metropolitan University. In the mid-1970s, he helped establish the journal *Food Policy*. He has also worked, on a long and short-term basis, in agricultural development projects in various parts of the world.

Christine Townend has written five books, including *Pulling The Wool* (Hale and Iremonger, Sydney, 1985), a devastating critique of Australian sheep farming. She founded Animal Liberation in Australia in 1976. Since 1990 she has lived in India, where she is a Managing Trustee of Help in Suffering Animal Shelter in Jaipur.

John Vidal has been the widely respected Environment Editor of *The Guardian* since1992. He is well known for his hard-hitting articles on key environmental issues. He recently published *McLibel – Burger Culture on Trial* (Macmillan, 1997), a vivid account of the famous libel case. He has twice won Environment Journalist of the Year in the national awards.

Mark Watts worked as a Local Government Officer and a Borough Councillor until 1994, when he was elected MEP for Kent East. He is Transport Spokesperson for the European Parliamentary Labour Party and is a Member of the Agriculture Committee of the European Parliament. He is Vice-President of the Council for Education in World Citizenship and a member of SERA, the Co-operative Party and CIWF.

Editor's Note

This book tackles a fundamental global concern – how can we feed ourselves in the coming millennium, at least in the coming century, in a way that provides healthy and plentiful food for all people and yet is gentle on animals and on the environment? If you eat meat, as I do, this book will almost certainly make you think twice about what you eat and the impact millions and millions of us following similar eating patterns has – both on the planet and the way we relate to each other. If you are a vegetarian or vegan you will find ample reasons to debate with your meat-eating friends about their habits. But for both groups the book will widen the horizons about the enormous impact the meat business is having globally – on people, domestic animals, wildlife and the environment – and the effects of new trade rules that could subvert other measures designed to produce sustainable livelihoods for people.

If to be human is also to be humane, then this book faces meat eaters with stark questions about the humanity of the current system, the forces driving it and the direction we are heading. Heading, that is, unless changes happen. Compassion in World Farming (CIWF) want readers to help make changes happen, to open up understanding and debate about the effects of the meat business and push for a different approach to farming.

This book developed out of CIWF's spring '98 conference – 'An Agriculture for the New Millennium – Animal Welfare, Poverty and Globalisation'. I reported on the conference for various print and radio outlets and I was surprised sometime later at Earthscan's suggestion that CIWF commission me to work on turning the conference papers into a book. I am not a vegetarian nor a member of CIWF but I too felt the issues raised at the conference deserve far wider debate than was possible with those who attended and believe this book should help achieve that.

Geoff Tansey
January 1999

Introduction

Joyce D'Silva

Before sitting down to write this introduction, I read through the fascinating array of chapters by this cornucopia of inspired authors and I found one statistic that kept popping up in chapter after chapter: 800 million people are not getting enough to eat.

Statistics like this are all too easy to dismiss, as not one of us can *imagine* 800 million people. But we can imagine just one person – possibly someone we know or love – and imagine them being undernourished, possibly even painfully hungry, day after day.

But statistics like this are horrifying and can reinforce our feelings of inadequacy – how can I make a difference?

I believe that this book will not only enable you to understand why so many of us are hungry – and so many overfed and badly fed – but will empower you to see practical ways in which you can help to make things better.

Another recurring theme in these pages is our exploitation of animals, in particular the animals we farm for food. You will find no animal liberation diatribe here, but an honest account of how we inflict immense and enduring suffering on billions of animals kept in factory farms to provide us with 'cheap' meat. You will discover how this cheapness is an illusion, how the intensive factory farming of these animals grossly distorts the world food supply and directly contributes to that undernourished individual and all the other 799,999,999 hungry people.

As this book makes clear, cheap meat takes food from the mouths of the poor; it uses in total more energy than it produces; it creates enormous environmental degradation and pollution; and it has turned the world's farmers into slaves of agribusiness or landless peasants filling the back streets of our urban conglomerates. José Lutzenberger, who has witnessed the obliteration of so much of Brazil's forests, describes the modern farmer as 'not much more than a tractor driver and poison sprayer'.

The dignity of the small farmers who knew their land and their animals intimately, who maintained soil fertility, rotated crops, used green manure and compost, used draught animals for energy, wind or water power for milling, and sold most of their produce at the local market, has all been largely destroyed.

Modern farming is big farming. It needs space for its monoculture crops so it rips out hedges and trees. It needs hybrids or genetically engineered crops for uniformity and yield, so it dismisses the vibrancy of biodiversity. It abandons natural soil fertility for addictive mineral fertilizers. It has taken the poison gas technology developed over two world wars and turned it loose on the insect world and indirectly onto us as consumers of multi-sprayed plants. It has taken drugs like antibiotics, originally developed to cure human infections, and applied them often indiscriminately to the animals in our factory farms. How else could thousands of chickens or pigs survive to slaughter weight in their crowded sheds, how else could their yields be maximized?

And in doing all this, control has been lost. Farmers of factory farmed chicken and pork are 'vertically integrated' into a system which dictates what food they feed the animals and sells it to them; it tells them when and where to slaughter and buys the end-product from them. Lettuce growers are told by supermarket chains how often and what to spray. Companies merge constantly so the agri-pharmaceutical giant becomes a transnational monster, controlling the genetically modified seeds and selling the herbicide to go with them.

Governments – democratic or otherwise – sign up to multilateral trade agreements like the General Agreement on Tariffs and Trade (GATT)/World Trade Organization (WTO) in the belief that free trade will bring its own divine equilibrium and balance. Yet common sense asks how the impoverished and debt-ridden can compete with the overabundance of the bank-rolled, well-watered Northern producers. Increasingly, governments seem so dependent for stability on the economic and financial structures that they have helped create, that their own power has been curtailed. They seem ever anxious to do the bidding of the transnational corporations or the bankers rather than listen to the concerns of the electorate.

Electorates are people – you and me. We all, or nearly all, want justice and fairness, not only for ourselves but for others, we want humane conditions for farm animals, we want safe and nutritious food, we want an agriculture which is gentle on the environment.

Yet the opposite happens and the true cost of our food and agriculture is, as Tim Lang succinctly puts it, 'dumped onto the environment, dumped onto the poor..., dumped onto health, dumped onto consumers'. And so we all lose out – as the forests of Brazil turn into soya bean horizons grown to provide cheap feed for the factory farmed animals of Europe, as the backyard pigs of China are rapidly multiplied and imprisoned in intensive units and a country which less than a decade ago was exporting grain now has to import it to feed its porcine millions, as debt-burdened African nations turn to producing exotic vegetables, fruits and flowers for sale in our trendy out-of-town supermarkets.

Links everywhere. Not forgetting our own health. With one in three Americans seriously overweight, and governments accused of colluding in dietary degradation, we all stand to benefit from a dietary revolution. Geoffrey Cannon presents the grim facts on cancer and diet and promulgates the healthier future of a primarily plant-based diet.

Dennis Avery, the archetypal agribusiness adherent, agrees that one way to save the world food situation is for five billion people to become vegan, although he thinks this an impossible goal to achieve.

Mark Gold, Christine Townend and Michael Fox speak longingly of the possibility of a utopian future where *ahimsa* – non-violence – rules in our relations with animals, in agriculture and between each other.

The professional politicians, Chris Mullin MP and Mark Watts MEP, see practical agendas for reform. Mark exposes the befuddled workings of the Common Agricultural Policy (CAP) which he says 'now constitutes one of the world's greatest manufactured disasters'. Chris appeals for a new approach to agriculture 'to restore morality to an industry which has lost its way'.

Atherton Martin, who has witnessed the power of big business which wanted to turn one third of the land in his homeland, Dominica, into a copper mine, warns as well of the 'enslavement of the *mind* by global information systems'.

Dennis Avery makes a valiant case for further intensification of arable and animal farming. He says only by tripling output can we save the wild lands from further intrusion. His is an argument that needs careful consideration, as it looks like being the policy of many governments for the next century.

But who will benefit from this intensification? Not the remaining small farmers, too poor to buy the patented, genetically modified seed; certainly not the animals forced to live in the mental and physical desert of the factory farm; probably not consumers either, who may retain their cholesterol-laden diets, now with the added risk element from gene-changed soya and corn.

Patrick Holden believes that the abuse of animals in industrial agriculture 'has incurred an enormous moral debt' – he is convinced that moving to animal welfare-friendly organic systems will help repay that debt and will bring abundant health and environmental benefits to us all.

Philip Lymbery describes in detail the kinds of abuse our factory farmed animals suffer, with misshapen bodies derived from extreme selective breeding and cage or crowd environments in which natural behaviour becomes impossible or is diverted into repetitive stereotypies. If you cannot root in the soil as a pig should, then you can always gnaw the bars of your crate. Janice Cox describes how Central and Eastern Europe are now caught at the crossroads between accepting investment in factory farming or taking serious steps to develop animal welfare laws.

But how can this cycle of deprivation be broken? Yes, the CAP can be reformed – we can subsidize good practice not bad – but there's still the GATT with its monolithic uniformity, its insensitivity to the needs of animals and environment. Peter Stevenson points out the absurdity of the tuna–dolphin rulings in which the US ban on imports of tuna caught in ways which killed or harmed dolphins was declared to be GATT-illegal.

It is not just GATT/WTO rulings – it's the *threat* of being ruled against which now makes governments reluctant to take bold reforming measures. Already we see the European Union (EU) dropping its ban on imports of fur from animals caught in cruel leghold traps, as well as its ban on the marketing of cosmetics tested on animals, because of fears of being GATT-incompatible. Truly this is scandalous.

John Vidal looks at the growing power of corporations in the food system and declares, 'What no human can do, a corporation can. It can turn one blind eye to suffering, another to natural justice. It cannot see common wealth, or culture, or tradition'.

Vandana Shiva, with her powerful background of research and on-hands campaigning, describes how a huge meat export industry has been jump-started in India by massive government subsidy, flying in the face of traditional sentiments of animal protection. Maneka Gandhi elaborates in detail on the suffering this has caused to the animals and the devastation it is causing to the environment, as the huge increase in goats has led to topsoil erosion. She says 37 per cent of India's agricultural land has now been diverted to grow crops to feed the animals for the meat export trade.

Again we hear about the powers of business and of multinational trade agreements affecting the lives of the poorest to the richest. As Vandana says, 'This rule of profits destroys the rule of compassion and justice'. In India she sees a clash between 'rule-based systems that protect life and defend the right of life of humans and all species on earth and a rule-based system that protects profits at any cost'.

Like so many other contributors to this book, Vandana has a sense of 'this amazing moment of history' when we seem to be at the crucial point of decision when, as she puts it, we can 'replace the rule of commerce enshrined in GATT, with new rules in which solidarity emerges between the rich and poor of the world and all species on this planet'. Back to this gentler way of living and being.

For these chapters are an extraordinary mix of the practical and the visionary, of those championing much needed political and legal reform to those who are convinced that long-term fundamental change can only come when we search our selves, our souls, our spirits, for an inner peace that can then reflect outwards in our actions. Michael Fox reminds us that we are all *humus* beings – he declares, 'Humility, humanity and humus are words that connect and ground us in the reality of our being'.

This book encompasses so much cutting edge thought, so much urgency for reform, so much knowledge and insight, it is truly inspirational. Possibly *the* book for millennial reading. I think I can guarantee it will enrich your life and, possibly, change it forever.

Part I
Conflicting Visions

There are very different measurements of the success of industrialized agriculture in feeding the world, and its capacity to continue to do so indefinitely. Here two strong critics and an equally strong proponent of intensive farming methods present their deeply divergent world views. Both perspectives may challenge your perceptions.

1 The Absurdity of Modern Agriculture: From Chemical Fertilizers and Agropoisons to Biotechnology

José A Lutzenberger and Melissa Halloway

In the controversy surrounding biotechnology and agriculture, the key issue is why and how agricultural production is increasingly dominated by giant corporations. The present, almost total, control of biotechnology by big business is the culmination of a process that has been going on for the last three quarters of a century.

Farming was invented some 10–15,000 years ago, and in the last 2–3000 years it evolved into many locally adapted, beautiful and sustainable peasant cultures around the world, especially in Europe, Asia, Mexico, Central America, the Andes and parts of Africa. From early colonization, American farmers, in spite of many disasters, such as the dust bowl, also developed beautiful farming systems, that were on their way to becoming sustainable. Many of these cultures were still intact right until the end of World War II. The few now remaining are being disrupted.

Industry has succeeded in successively appropriating from the farmers more and more of their activities, taking away from them all that enables it to reap secure profits and leaving the risks with the farmers – the risk of bad harvests from bad weather and the risk of losing money due to growing dependency on inputs that have to be purchased at rising prices and having to sell their produce at continuously falling prices.

The conventional argument in favour of the methods of modern agriculture is that they are the only efficient way of solving the problem of world hunger and of feeding the masses that are still to come with the population explosion. But this is an illusion. Traditional peasant methods could be improved with today's scientific knowledge of how plants grow, of soil structure, soil chemistry and soil life, as well as of plant metabolism and so on. But the improvement need not be in the direction of gigantic monocultures, highly mechanized and with all the paraphernalia of commercial fertilizers and synthetic poisons, with agricultural produce being transported all around the globe. Big monoculture was an invention of colonialism. The colonial powers could not extract much from the traditional peasantry with their highly diversified crops for subsistence and for local and regional markets. They wanted great amounts of cotton, sugar, coffee, tea, cocoa and the like. This led to the uprooting of millions of people and was also at the root of the slave trade from Africa to the Americas, one of the greatest calamities of human history.

However, the fundamental problem with modern agriculture is that it is not sustainable. Even if it were as productive as is claimed, disaster will only be postponed and will then be much worse. If we are to feed the growing masses – of course we will have to find ways of bringing our numbers under control – then we must develop methods of sustainable agricultural production.

With very few exceptions, traditional peasants had developed sustainable methods. Chinese farmers, for example, for 3,000 years obtained high productivity from their soils without compromising fertility. On the contrary, they built up and maintained maximum soil fertility. Modern regenerative farmers are learning to become increasingly sustainable with optimum yields and locally adapted methods, while recovering and maintaining biodiversity in their cultivars and in the surrounding landscape. Let us call them regenerative farmers, not biological, organic or alternative. When we deal with life, everything, whether good or bad, is biological, is organic, even mass slaughter. Alternative only means different, it could be worse. But regenerative means that it regenerates what had been lost or destroyed.

Modern agriculture has stepped outside of the logic of natural living systems. All natural ecosystems have automatic internal feedbacks that, from the very beginning, such as when a new barren piece of land, say the slope of a volcano, is conquered, make environmental conditions improve until a climax of maximum sustainable biological activity is achieved. Our modern agricultural ecosystems do the exact opposite, we then impose feedbacks (agri-chemistry) that increasingly degrade the environment and impoverish biodiversity.

Unfortunately, modern farming succeeds by mining the soil and replacing lost fertility with imported nutrients. Commercial fertilizers, such as phosphates, come from mines that will soon be exhausted. Potash mines are more plentiful, but nitrogen, the most important element in modern agricultural productivity, even though it comes from the atmosphere, a virtually inexhaustible source, is obtained in the Haber-Bosch ammonia synthesis, a process that consumes enormous amounts of energy, mostly energy from fossil fuels. Even when it is energy from hydropower, it is electricity that could be saving fossil fuels somewhere else. All the other inputs, such as agripoisons and increasingly heavy machinery, are also highly energy intensive.

But agriculture, if we look at it from a holistic, ecological perspective, is a scheme for harvesting solar energy via photosynthesis. Whereas all forms of traditional agriculture had a positive energy balance, modern agriculture has perverted even this fundamental aspect. Most of it has become a net consumer of energy. Almost all of its supposedly highly productive operations require more fossil energy, on the whole, than is contained in their produce. To use a fitting metaphor, it has become like an oil well where the engine that drives the pump consumes more fuel than it can bring up. This kind of operation can only survive with subsidies.

Modern agriculture is said to be so efficient that only about 2 per cent of the population can feed the whole population. Until the turn of the century, in Europe, in the US and in most countries, about 60 per cent of the population worked on the land. By the end of World War II it was still close to 40 per cent. Today, in the US less than 2 per cent of the population are farmers. In most European countries, the figure is approaching 2 per cent, as farmers are still forced to give up. Now, when it is said that in a modern economy only 2 per cent of the people can feed the whole population, as against 60 or 40 per cent in the past, that is either an illusion or a lie. It is based on a wrong comparison.

In the economy as a whole, the old peasant farming system was a system of production and distribution of food that also produced its own inputs. Soil fertility was maintained with dung, crop rotation, companion planting, green manure, compost, mulching and fallow; the seed were selected from the best of each crop; draught animals supplied the energy; in the mills it was wind or water power. It was all solar energy. Most of the farmer's produce was delivered almost into the hands of the consumer on the weekly market.

But the modern farmer is only a little cog in an enormous techno-bureaucratic infrastructure that even requires special legislation and heavy subsidies. Compared to his predecessors who did almost everything that had to do with food production, processing and distribution, he is not much more than a tractor driver and poison sprayer.

After World War II, when Germany was totally devastated, the Marshall Plan helped recovery. More important, however, was the ability of city people to swarm out into the country to 'hamster', that is, barter anything of value – a watch, a ring, a piano – for some food. The farmers had food – they had grain, beans, potatoes, vegetables, fruits, milk, cheese, chicken, geese, and much more.

It would not take a war now to put European farmers in a position where they would have to 'hamster'. But then, where? Not a single bomb need fall. Collapses in energy, in transportation, especially importation of mineral fertilizers and cattle feed, in the banking system and even in communication and the computer networks, could do it. Amazingly, the military do not seem to be concerned. Fundamentally, national security depends on healthy, sustainable farming.

Today's system of food (including fibre and a few other non-food items) production and distribution begins in the oil fields and all kinds of mines for metal and other raw materials, goes through the refineries, steel and aluminium smelters, etc, the chemical industry, the machine industry, the banking system, the all-embracing transportation system (mostly fossil fuel consuming), computer, supermarket, packaging industry and a whole new complex of industries that hardly existed in the past – the food manipulating industry that rather deserves the name food denaturing and contaminating (with additives and residues of agripoisons) industry. If we want to compare today's farmer with traditional peasants, then all the working hours in the above mentioned industries and a few others, as well as some services, such as junk food joints, to the extent that they directly or indirectly contribute to the production, manipulation and distribution of food, must be added up. This should even include the working hours that correspond to the money that has to be earned in order to pay for the taxes that pay for the subsidies. Significantly, the biggest chunk of the subsidies goes not to the farmer but to the industrial complex. The farmer is always left on the brink of foreclosure.

A complete balance of this type would certainly show that today, in a modern economy, also about 40 per cent or more of all the working hours go into the production, handling and distribution of food. Unfortunately, today's conventional economists, those to whom our governments listen, in their non-holistic worldview, list the tractor and combine factory with the machine industry, the chemical fertilizer and agripoison factories with the chemical industry and so on, as if they had nothing to do with food.

What we have then, with a few exceptions, is a redistribution of tasks and certain forms of concentration of power in big business, not more efficiency in agriculture.

ILLUSORY GAINS

Frequently, the modern food production and distribution system is not only no more productive with manpower but also is no more productive in yield per acre. In many cases, such as in intensive animal rearing, it is even destructive. It destroys more food than it produces.

In Southern Brazil, during the past half century the great subtropical forest of the Uruguay valley was completely obliterated, leaving only a few small relics. The forest was cleared and burned, destroying the timber, to give way for soya bean monocultures. This was not done to relieve the problem of hunger in the poor regions of Brazil but to enrich a minority (people with no agricultural tradition) through soya exports to the European Community for cattle feed. These soya bean plantations are among the most modern anywhere – large, highly mechanized and with the usual chemical inputs. In Brazil's subtropical climate, the farmer has the added advantage of being able to grow wheat, barley, rye or oats or to make hay and silage in winter on the same soil. Compared to what our peasant farmers did on similar soils, productivity is low, seldom more than three tonnes of grain (total, summer and winter) per hectare. The peasant, who produced to feed the local population, easily produced 15 tonnes of food per hectare, diversifying with manioc, sweet potatoes, Irish potatoes, sugar cane and grains, plus vegetables, grapes and all kinds of fruit, hay and silage for his cattle, and he had chickens and pigs.

Despite this reality, official agricultural policies have always supported the big guy at the expense of the peasant. Hundreds of thousands of peasants had to give up. They either went to the cities, often to the slums, or migrated further north all the way to the Amazon rain forest. Tremendous devastation was caused with World Bank money in the State of Rondônia, and the small farmers who were settled there, not knowing how to farm in the tropics and with no help, are mostly failing, leaving devastation behind while new forest is cleared further on. In Central Brazil, the Cerrado, the South American equivalent of the African savannah, is now being almost totally destroyed for still more soya bean plantations, one of them over 100,000 hectares in one contiguous piece. In its biodiversity, the Cerrado is as precious as the tropical rainforest, even more so in some parts.

Some also argue that the Indian peasants in Chiapas, Mexico, who are now fighting for their survival by rebelling against NAFTA, the North American Free Trade Agreement, are backward, that they produce only two tonnes of maize per hectare as against six on modern Mexican plantations. But this is only part of the picture; the modern plantation produces six tonnes per hectare and that is all. The Indian grows a mixed crop – among his corn stalks, that also serve as support for climbing beans, he grows squash and pumpkins, sweet potatoes, tomatoes and all sorts of vegetables, fruit and medicinal herbs. From the same hectare, he also feeds his cattle and chickens. He easily produces more than 15 tonnes of food per hectare and all without commercial fertilizers or pesticides and no assistance from banks or governments or transnational corporations.

The uprooting of people such as these is the continuation of one of the greatest disasters of modern times. When they land in the slums of the cities they will have to buy food grown on farms that are less productive than they were. On balance, there is then less food and more people to feed. Often their land is then taken over by extensive cattle ranchers who seldom produce more than 50kg of meat per ha/year. Hundreds of similar stories could be told. In the case of Chiapas, people in every valley spoke a different language, had a different culture. On top of all the personal calamities, when the landscape is cleared of its traditional peasants we have cultural genocide!

MORE MEAT, LESS FOOD

In mass animal rearing for meat and eggs, the methods are downright destructive. Much more food for humans is destroyed than is produced. The chickens in their sad concentration camps or egg factories, euphemistically called 'chicken farms', are fed 'scientifically balanced' rations consisting of cereal grains, soybeans, palm oil cake or tapioca, often with fish meal. We know cases in Brazil where chicken feed contains powdered milk from the EU. This puts them in competition with humans. We feed them from our crops. This is a total absurdity if the aim is to contribute to solving the problem of world hunger. In traditional agriculture, chickens ate insects, worms, manure, herbs and grasses and refuse from kitchen and crops, thus increasing the carrying capacity of the farmers' land for humans. Now they diminish it.

The ratio of transformation of feed to human food is close to 20:1. Half the weight of the living animal – feathers, bones, intestines – is not consumed by us. Moreover the concentrated feed rations are dried with a high input of energy to a maximum of 12 per cent of water while meat is up to 80 per cent water. In the feeding barn, the most efficient operations use about 2.2kg of ration to obtain 1kg of living weight of chicken, half of which is human food. So 2.2:1 becomes 4.4:1. If we now correct for water content: (4.4 x 0.88 for feed and 1 x 0.2 for meat) we get 3.87:0.2, which equals 19.36:1.

More recently, some chicken companies have 'improved' the ratio somewhat by including in the rations offal from the chicken's predecessors in the slaughterhouse, thus forcing them to cannibalism. There is still one more absurdity: the 'scientifically balanced' rations contain nothing green. It is the same for pigs. But chickens and pigs are voracious eaters of herbs, grasses, fruits, nuts, and roots. In our experiments with sustainable farming we also feed them waterplants, with great success – healthy animals, no antibiotics, no drugs, no veterinarians.

And finally, in the chicken concentration camps and egg factories, as well as in the modern pig dungeons, the poor creatures live under conditions of extreme stress.

It is time to expose the lie that only agriculture as promoted by technocracy can save humankind from starvation. The opposite is true.

We need a new form of economic accounting that, when it adds up what is called 'productivity' or 'progress' in farming, also deducts all the costs: the human calamities, the environmental devastation, the loss of biological diversity in the landscape and the even more tremendous loss of biodiversity in our cultivars. This last aspect will be enormously aggravated with biotechnology as handled by big business.

Concentrating control

Brazil is a big exporter of chicken meat, mostly to the Middle East and Japan. It is easy to see how the destructive methods of industrial chicken operations developed. In Southern Brazil, it began in very simple schemes, where small, individual entrepreneurs confined chickens in a barn and fed them maize. The system coalesced and grew to the point where, today, there are half a dozen very large companies and a few small ones. The big slaughterhouses may kill and process up to hundreds of thousands of chickens a day. They operate according to rules, established by them, that they call 'vertical integration'. The 'producer' signs a contract where he accepts buying all his inputs, hatched chickens, feed and drugs, from the company. Even if he is a farmer and happens to have plenty of grain, he is not allowed to feed it to his chickens. He must buy the ready made ration, but he can sell his maize to the ration factory that belongs to the same company that owns the slaughterhouse and that also owns the hatcheries that produce the chicks. These operate a different type of chicken concentration camp where the prisoners are cocks and hens, one cock to ten hens. The hens are not in small cages as in the egg factories, they can move freely within the barn and jump into ample nests for laying. (In the conveyer belt operations of the egg factories, called batteries, the poor hens sit, three to a cage, on a wire grid and the eggs roll out.) The chicks produced in these hatcheries are not traditional chicken races anymore, they are registered brands and they are hybrid chickens. Just like hybrid maize, they cannot be reproduced true to race.

After buying all his inputs from the company he signed his contract with, the farmer can only sell to the same company. He is not even allowed to sell to one of its competitors, and they would not take it. So he may live with the illusion that he is a self-employed small entrepreneur, but his real situation is that of a worker with unlimited working hours, no weekends, no holidays and no vacations, and he has to pay for his own social security. If the big company used hired workers, they could not do it, it would be too expensive and too risky. This way they leave all the risks with the producer: loss through disease plus additional costs with drugs and antibiotics, heat stroke (a common disaster during hot summer days, when hundreds or thousands of chickens die in the crammed and badly ventilated conditions of the barns) and loss during transportation. The chickens that die in the company's trucks on the way to the slaughterhouse are also discounted. The farmer's profits are also constantly shrinking as input prices increase and proceeds from his sales fall. The producer's margin is so tight that, even if everything goes well, but if he has to feed his animals a few more extra days, his profit may evaporate or even turn into a loss. This is a common occurrence. The slaughterhouse schedules its trips for the collection of the ready chickens according to its own convenience, not the producers'. And if the company receives a windfall thanks to better prices in the export markets, it does not share it with the producer.

These chicken concentration camps have nothing to do with higher productivity to help save humankind from starvation – in fact, they contribute to the problem – but they concentrate capital and power by creating dependency.

These methods were not invented by farmers. A farmer in a healthy peasant culture would never feed masses of his grain to chickens, unless it were rotten grain, and so isolate them from their natural food source, thus wasting part of his soil's carrying capacity for humans, while destroying part of his harvest. These methods are also not the result of a concerted conspiracy by technocracy. Such schemes grow naturally from an initial 'seed' that may have had a completely different intention. In this case, as was the case with agrichemistry, it was the war effort. The conspiracy grew organically over time. During the last World War the American government initiated the subsidy system for grain production, which led to enormous surpluses. So, agricultural authorities looked for 'non-human uses' for grain. 'Vertical integration' is only the momentary stage in the process of concentration of power. Soon they will find ways of banning – by special legislation – the rearing of free roaming chickens by independent farmers. They already tried, unsuccessfully, but they succeeded in making it very difficult for small farmers to sell eggs from such chickens on the open market.

There was also no conspiracy with hybrid maize at the beginning, it came later. Plant breeders discovered that crossing two super-pure strains of maize – strains obtained by inbreeding for eight to ten generations – produced plants of high productivity and perfect uniformity. For them it must have been a disappointment when it turned out that the cross was not stable. Upon reseeding it 'Mendeled out', so to say, according to Mendel's law of desegregation. The new crop was chaotic – tall stalks, short stalks, one cob, many cobs, different colour, shape and quality of grain. But, from the point of view of the seed merchant, it was a true advantage! Now the farmer could not save his own seed, he had to buy new seed every year. The merchant did not even need the protection of a patent.

Fortunately for most crops, especially grains such as wheat, barley, rye and oats, this type of hybridization is not yet economically feasible for the breeders. They are trying with every cultivar they can get their hands on. But it works with chickens. In Southern Brazil we have to have an association to preserve traditional chicken races. Most are now threatened with extinction. Some are already gone. Only the registered brands of hybrid chicken are not threatened. As for maize, almost all the traditional varieties are gone. If a farmer wants to grow one of them he gets no credit from the bank. Only the 'registered' varieties are accepted.

Now, direct genetic manipulation, called biotechnology, that operates at the chromosome level, gives breeders a shortcut in the direction of taking control of cultivars away from the farmer. But, since most of the products of direct gene manipulation do not desegregate in reproduction, the breeders now need patents.

FROM AGRICHEMISTRY TO BIOTECHNOLOGY

Until the end of the 1940s, agricultural research looked for biological solutions. The perspective was ecological, though there was hardly any talk of ecology. Had this trend been allowed to continue we would today have many forms of locally-adapted highly productive, sustainable agriculture. But, beginning in the 1950s, the chemical industry managed to fix a new paradigm – in the schools, in agricultural research and extension. I call it the NPK+P Paradigm. NPK stands for Nitrogen, Phosphorus, Potash (Potassium), the second P is for pesticide or rather poison.

Commercial fertilizers became big business after World War I. The Allied blockade at the start of the war cut the Germans off from Chilean nitrate, essential for the production of explosives. The Haber–Bosch process for the fixation of nitrogen from the air was known but had not been exploited commercially yet. The Germans set up enormous production capacities and managed to fight for four years. Without this process, World War I would not have really developed, there would have been no Treaty of Versailles, therefore, no Hitler…! It is staggering how one technology can alter the course of history.

When the war was over, there were enormous stocks and production capacities but there was no more market for explosives. Industry then decided to push nitrogen fertilizers onto agriculture. Up until then, farmers were quite content with their organic methods of maintaining and increasing soil fertility. Chilean saltpeter and guano were used in a very limited way, only on special crops, mainly in intensive gardening. Nitrogen fertilizers in the form of concentrated, almost pure salts, the nitrates and ammonia fertilizers, are kind of addictive – the more you use the more you have to use. They soon became very big business. So the industry developed a complete spectrum, including phosphorus, potash, calcium, the microelements, even in the form of complex salts, applied in granulated form, sometimes applied from a plane.

World War II gave a big push to a small, almost insignificant pesticide industry, and really got it started on a big scale. Today, dozens of billions of dollars worth of poisons are spread all over the planet – $28.5bn according to the last figures from GIFAP (Groupement International des Fabricants de Pesticides). During World War I poison gas was used only once, with devastating effect on both sides, therefore never used again. During World War II, no gases were applied in battle, but a lot of research was conducted. One major company developed the phosphoric acid esters. After the war, they had large production capacities and stocks and they decided that what kills people should also kill insects. They made new formulations of the stuff and sold it as insecticide.

The chemical DDT was known as a laboratory curiosity. When it was discovered that it killed insects without, apparently, affecting people, the American armed forces, who were suffering from malaria in the Pacific while fighting the Japanese, were alerted. They used it in a totally ill-considered way – convinced as they were of its harmlessness – blanket spraying it over whole landscapes and even into homes and on people's clothes.

Shortly before the end of the war in the Pacific an American freighter was on its way to Manila with a load of potent plant killers of the 2,4-D and 2,4,5-T group. The intention was to starve the Japanese by destroying their crops by spraying the plant poison from the air. It was too late. The boat was ordered back before it arrived. Another group of Americans had dropped the atom bombs on Hiroshima and Nagasaki, a terrible story everybody knows, and the Japanese signed the armistice. Again, the same story: large production capacities, enormous stocks with no buyer. The stuff was reformulated as 'herbicide' and unloaded on the farmers. Later, during the Vietnam war, the American armed forces ruthlessly sprayed what they called 'Agent Orange' (and other colours) on millions of hectares of tropical forest, pretending it to be only a 'defoliant' to make the enemy forces visible. In fact, these formulations contained high concentrations of 2,4,5-D that totally destroyed the forest.

Industry, wanting to preserve into peacetime what had become big business during wars, managed almost completely to take over agricultural research and to redirect it to its own aims. It also coopted official research and extension as well as schools and, lobbying for adequate legislation or regulation and setting up banking schemes for (apparently) easy credit, put the farmer in a position where there were hardly any alternatives left. Today, the agrichemical paradigm is accepted almost without question in the agricultural schools, in research and extension. The majority of farmers, even those who are uprooted, believe in it and often blame themselves for their incapacity to cope.

All this came about not as a deliberate conspiracy by evil-minded people, it just developed and structured itself from opportunism to opportunism. To the extent that a new technique, process or regulation gave somebody or some institution advantage, that technology was pushed and ideologically consolidated. Alternatives that did not fit in with the growing power structures were fought, ignored or demoralized.

Now, in the case of biotechnology in agriculture controlled by big transnational corporations, it seems that we do have a true conspiracy and that the damages will be much more irreversible than what we had up to now.

The main issue here is not so much whether our food will become of inferior quality and even harmful – even though it may – but, again, it is a question of adding up still more structures of dependency, of domination over remaining farmers and limitation of choice for the consumer.

The fantastic diversity of cultivars we had and still have today after the tremendous loss caused by the 'Green Revolution' during the last few decades, is the result of conscious and unconscious selection by peasants themselves through the centuries and millennia. Just think of the family of *Cruciferae* – cabbage, Chinese cabbage, radish, turnip, mustard, cauliflower, broccoli, colsa and many others. None of these farmers ever asked for patents, registration or certification.

Now industries, such as Monsanto, want us to accept their genetic manipulations from that pre-existing wealth, such as 'Roundup-ready soybeans', with the argument that they are only continuing and accelerating that process, thus contributing to solving the problem of feeding humankind. They even insist that there is no other way. They should know there are alternatives; better, healthier, cheaper ones.

Everybody knows that agriculture must find ways of getting away from poisons. We have all the knowledge necessary. Thousands of organic farmers all over the world are proof of it. With herbicide resistant cultivars the industry wants to sell a package – seed plus herbicide – forcing the farmer to use a herbicide, even if he does not need it, and to use their herbicide. In the case of cultivars with the infamous 'terminator gene' the conspiracy is even more obvious. With this kind of seed they do not even have to go to the trouble of applying for patents. All this has nothing to do with increased productivity, it is the culmination of the on-going process of disappropriating farmers, to turn the surviving ones into mere appendices of industry. It will aggravate uprooting, social disruption, environmental devastation and loss of biodiversity in nature and in our cultivars. It will aggravate the problem of hunger.

2 Intensive Farming and Biotechnology: Saving People and Wildlife in the 21st Century

Dennis T Avery

The 21st century will present the world with the greatest agricultural challenge in history. A larger, more affluent human population will demand three times as much food and fibre as farmers currently produce. The key question for both people and wildlife is how to produce this additional farm output without clearing all the forests and wildlands for additional crops and livestock.

The urgent competition between people and wildlife for land has been difficult to understand in Western Europe, where a foolish farm policy has given the illusion of food surpluses for the past 30 years. But the world never had a farm surplus, and now the CAP is in an openly declared state of collapse. The farm price supports will be eliminated eventually – along with the illusion of surplus, the cropland set aside, and probably the subsidies for organic farming.

The affluent, insulated community of European vegetarians and animal rights activists is about to be drawn into the real world of the 21st century. They will face challenges and opportunities beyond their previous imaginings. Most especially, they must confront the huge realities: the world must create five billion vegans in the next several decades, or triple its total farm output without using more land. The prospects for creating all those vegans are poor.

Thanks to biotechnology, the prospects for tripling its crop yields are much better. In fact, biotechnology may be the only compassionate answer to the world food challenge in the 21st century – for poor people, for children, and for the billions of wild creatures on the planet.

POPULATION RESTABILIZING

The world's human population will peak at about 8.5 billion, about the year 2035, up more than 40 per cent from today. That is good news for the environment, compared to past forecasts of 15 or 50 billion humans. The birth rates in developing countries have already come down radically, from 6.5 per woman in 1965 to only 3.1 currently. Stability is 2.1 births per woman, so the poorest countries have come 80 per cent of the way to population stability in one generation.

Population growth is tapering off essentially because of affluence, and the broader knowledge that comes with it. The rich countries are already below replacement, at 1.7 births per woman, and the rest of the world is headed there as it becomes more urban and industrial.

Affluence is growing primarily because of spreading technology and trade. It is driving economic growth rates in the developing countries that are twice as high as the recent growth rates in the 'rich' countries. We can expect economic growth to continue, strongly, as long as the WTO continues to stimulate the reduction of trade barriers.

More affluent people will demand much more food and fibre. Compassion does not allow us to achieve otherwise compassionate goals by leaving small children hungry or poor people malnourished. We must find ways to produce more food.

LITTLE PROGRESS TOWARD A VEGAN WORLD

The day may come when the people of the world will give up eating meat. They may even give up eating milk and eggs. But there is little evidence of it in today's more affluent societies.

A recent US survey for the *Vegetarian Times* indicated that 7 per cent of Americans consider themselves vegetarians – but two thirds of those 'vegetarians' eat meat regularly, and 40 per cent eat red meat regularly. Only about 0.2 per cent of Americans are vegans, forgoing all of the expensive, high-quality calories from livestock sources.

Saving the wildlands with human dietary change might require that 70 per cent of Americans be vegan – and that will have to be achieved without restricting the meat and milk intake of youth. That might be injurious to their health. In other words, we might have to convert virtually every adult to vegan diets, after letting them taste meat and milk as they grow up. I cannot imagine a tougher challenge.

Meanwhile, the demand for meat in the world has been rising by more than five million tonnes per year. The biggest surge in demand has occurred in China, where meat consumption in 1998 is forecast at 63 million tonnes, up from 29 million tonnes as recently as 1991. In India, milk consumption has doubled since 1980, to 65 million tonnes, and milk supplies are still short of meeting demand. Poultry and egg demand has been rising by 20 per cent annually in Indonesia.

Today, there is no major visible global trend toward vegetarian or vegan diets. Nor is there any major global campaign that seems likely to produce such a trend. I am not even aware of the funding to support such a campaign, which would need at least a billion dollars to be credible as a worldwide agent of change.

Thus, in the critical effort to preserve the world's wildlands and biodiversity, we dare not count on a change in human dietary habits to ease the challenge facing world agriculture. Instead, we must supply at least three times the farm output for 2040 as the world's farmers are supplying today. Agriculture already takes 37 per cent of the world's land area – the cities take 1.5 per cent. Thus, we must triple the world's crop and livestock productivity per acre in the next 50 years, or see most of the world's wildlands destroyed. And we cannot, with compassion, simply accept the destruction of the world's wildlife.

The Organic Constraints

Many people are demanding that we make the agricultural challenge even more difficult by restricting farmers to organic inputs. The manager of a large farm in the UK – which farms part of its land organically and part with chemical support – says that he gets about half as much yield from his organic acres as from the rest of the farm. The editor of the *American Journal of Alternative Agriculture* recently told me that the average US organic farm has about one third of its land in green manure crops and fallow to make up for the lack of chemical nitrogen that mainstream farmers take from the air.

If organic farming yields are half as high as mainstream yields, then meeting the 21st century food challenge would require organic farmers' yields to be raised sixfold in the next decade or two. If organic yields are two thirds as high, then organic farmers would have to increase their yields by only about fourfold to save the wildlands. Based on current knowledge, it is difficult to see how such huge increases could be achieved without important new knowledge being discovered and applied.

Worldwide, sacrificing one third of our farms to green manure crops might mean ploughing another five million square miles of wildlife to get 'natural' nitrogen.

DESPERATE REGIONS: THE FOOD GAPS IN AFRICA AND ASIA

The world's traditional pattern of agriculture has always featured small farmers supplying nearby consumers with seasonal fresh foods. Unfortunately, tripling the world's farm output on this model for the 21st century would likely mean sacrificing at least half of the world's tropical forests to slash-and-burn farming. Such farming is cheap and effective for low levels of population density. But Africa's population is projected to grow from 200 million to at least 400 million. Asia's population will rise from 3 billion to 4 billion. Neither region is yet providing its consumers with the high quality diets they increasingly demand and can afford, so the developing countries must accommodate not only a huge further increase in population, but an even larger increase in dietary quality.

Asia, for example, is providing only about 17 grammes of animal protein per capita per day. Europeans and North Americans consume 65 to 70 grammes per day, and the Japanese 60 grammes, up from 15. By 2030, we can expect 4 billion Asians to demand at least 55 grammes per day of animal protein – more than a fourfold increase!

India is getting one third of the fodder for 400 million dairy animals by literally stealing leaves and branches from its forests. Indonesia is clearing millions of hectares of tropical forest to grow low-yield crops of rice, corn and soya beans, and to provide poor quality cattle pasture. Africa has already dangerously shortened its bush fallow periods, from the optimal 15–20 years down to as little as 2–3 years in some regions. It cannot support the expanded population and rising expectations

None of this is environmentally sustainable. The world must have still higher yields of crops and livestock, and free trade, or it will lose most of its tropical forests, and perhaps three fourths of its 30 million wildlife species.

BIOTECHNOLOGY OFFERS HOPE IN THE 21ST CENTURY?

Much of the productive power of nitrogen and hybrid seeds has already been applied to get today's farm output. Tripling yields again will require us to apply more knowledge, more effectively.

Biotechnology seems to be the most promising way to ease the land conflict between people and wildlife in the 21st century. Biotechnology is the big new knowledge breakthrough that is just beginning to be applied to agriculture. It apparently has more conservation potential than any agricultural technology in human history. For example:

- Two researchers in Mexico have found a way to unlock the productivity of billions of hectares of acid-soil lands in the tropics. The acidity cuts crop yields by up to 80 per cent, on 30–40 per cent of the world's arable land, most of it in the tropics. Huge tracts of otherwise good land in Brazil and Zaire have simply been left unused, growing only stunted brush and poor quality grasses. But a gene from a bacterium has given crop plants (tobacco and papaya) the ability to secrete citric acid from their roots. (This is a success strategy used by some of the wild plants growing on the acid soils.) Apparently, the new biotechnological intervention will overcome much of the 'tropical disadvantage' which has kept regions like Central Africa and South Asia so poor for so long.
- Genes from wild relatives of our crop plants appear to be one of the most promising avenues for achieving safe, sustainable yield gains for the 21st century. Scientists have gathered hundreds of thousands of such wild relatives for the world's gene banks. However, these wild relations are too different from the crop plants to cross-breed. The wild-relative genes can only be used through biotechnology. But what promise they contain! Researchers from Cornell University have recently used wild-relative genes to get a 50 per cent increase in yields of tomatoes! (Tomato yields in standard cross-breeding programmes have recently been rising by only about 1 per cent per year.)
- The same Cornell research team inserted two promising wild-relative genes into the top-yielding Chinese rice hybrids. Each of the new genes produced a 17 per cent yield gain. Together, they offer the world's rice breeders a sudden 20–40 per cent increase in rice yields. It is no accident that China has just announced a

new rice variety that yields 13.5 tonnes in test plots – more than double that nation's 6 tonnes national average yield.

- In milk production, a genetically engineered copy of the cow's natural growth hormone, bovine somatotrophin (BST), is being widely used in many countries because it lowers milk production costs. It provides a 10 per cent increase in dairy feed conversion efficiency. For a country such as India, which has enormous unmet milk demand and a severe feed shortage, the bovine growth hormone should be a powerful conservation tool. It should mean that far fewer of India's cattle suffer hunger and malnutrition. Bovine growth hormone seems triply compassionate, for cows, children and wildlife.
- The new 'conservation tillage' farming systems that are currently sweeping across the world's best farmlands also bring enormous benefits. They are radically cutting soil erosion losses, increasing the health of soils and sub-soil biota, and raising crop yields higher than ever before. These conservation farming systems depend on herbicides rather than ploughing to control weeds. In the American Corn Belt, they are cutting soil erosion by 65–95 per cent. In the Cerrados Plateau of Brazil, they are permitting farmers to successfully crop the rolling, erosive soils of the Cerrados for the first time. In the rice countries of Asia, no-till farming lets poor peasant farmers prepare their rice paddies in 14 days instead of 60 days – making room for a third rice crop in countries such as Indonesia.

These are all examples of 'high-yield conservation'. Since 1950, the rising yields of the Green Revolution have permitted farmers all over the world to triple their yields (and more) on the world's best farmland. That is permitting the world to feed better diets to twice as many people, without taking any more land for farming (except in Africa). Without the yield gains, we would already have had to plough another 15 million square miles of wildlands – equal to the total land area of Europe, the US and South America.

I cannot say there will be no risks from biotechnology, to either people or the environment. I can say with confidence that without biotechnology for the agriculture of the 21st century, we will certainly destroy millions of square miles of wildlands.

Compassion and farming in the 21st century

I applaud the efforts of voluntary groups to conserve wildlife. I also applaud the long history of concern for ensuring the humane and compassionate treatment for domestic animals. However, until and unless the animal welfare movement can generate a far more successful campaign for vegan diets than the world has seen to this day, the most compassionate approach to agriculture may be what we, at the Hudson Institute, call 'high-yield conservation' – higher yield crops; higher yield pigs, chickens and cattle; higher efficiency irrigation; and higher yield tree plantations.

Another vital element of high yield conservation will be the confinement – or intensive – production of cattle, hogs and poultry. If world demand expands from one billion pigs to three billion, and from 13 billion chickens to 50 billion, the additional birds and animals must be raised in carefully managed confinement facilities.

If the US raised its chickens today on free range, it would mean taking from wildlife a land area the size of the state of New Jersey. It would mean much higher death rates for the birds, from both the pecking order and the diseases that are spread more readily in free-range birds. If hundreds of millions of pigs were raised on free range, it would mean not only millions of square miles of wildlands converted to pasture, but also massive soil erosion as the pigs rooted and wallowed. They would also face higher disease and death rates.

Recently, the US State of North Carolina has become one of America's pig production centres. When its State University did an intensive survey of those pig farms, it found about 5 per cent of them poorly managed, with manure and urine escaping into the environment, and with high rates of stress and disease for the animals. The offending farms, in virtually all cases, were the smaller, older, traditional pig farms. The big 'factory farms' which have drawn so much criticism were treating their animals the best!

In the years ahead, Western Europe's farmers will have to supply cereals, fruits, vegetables, meat, milk, and processed foods in direct competition with the other farmers of the world. There will be no EU preference, and no price subsidies. The rights of the domestic animals in Western Europe will have to be safeguarded despite intense global price competition, and virtually without subsidies. It will be an enormous challenge for the animal welfare movement.

Compassion and good judgment should guide an agriculture for the 21st century that features still higher yields and still higher levels of caring and compassion – for poor people, for children, for animals and for the environment. However, without higher yields from our crops, livestock and poultry, the world will descend into the most vicious competition for land ever seen. Hungry children will compete directly with the habitat of the elephant, the lowland gorilla and the Bengal tiger. Domestic animals will suffer malnutrition and privation on a scale never seen before. Wildlands which have been successfully preserved through the biggest surge in population and food demand the world has ever seen will be sacrificed by the millions of square miles to the plough of the low yield farmer. More and more domestic livestock and poultry will be raised by people less and less able to be concerned for them.

Compassion demands we avoid that.

Part II
Whose Path to Follow?

Whichever path we take, it should be underpinned by clear values. The choices facing US and Indian agriculture, that will affect our food future and the kind of environment we live in, are explored in Chapters 3 and 4. The question of who is making the key choices and where these are leading us is addressed by John Vidal in Chapter 5.

3 American Agriculture's Ethical Crossroads

Michael W Fox

The cattle of the rich steal the bread of the poor
M K Gandhi

American agribusiness claims to feed the hungry world and produce more food that is safe, nutritious and cheap than any other country. These and other myths are promoted by overcapitalized, petrochemical-based food industries and lack a bioethical basis. In fact, current agribusiness developments and their consequences in the US and abroad amount to a collision between American style industrial agriculture and reality. Increasing soil and water conservation problems and animal waste disposal problems, and environmental, food safety and quality concerns, coupled with future shortages of such key agricultural inputs as phosphates and fossil fuels, mean conventional agriculture in the US is at a crossroads. Current agricultural practices cannot continue along the same path. For example:

- The US Department of Agriculture (USDA) is torn between its allegiance to the livestock industry and the public interest. It tells people to eat less fat and more 'beneficial' fruits, vegetables and high fibre cereals but says nothing about reducing their consumption of animal products. Yet many studies now indicate this will help reduce the incidences of cancer, obesity and a host of other diet-related diseases.
- Manure runoff is linked with human health problems, including short-term memory loss in and around the East Coast, and massive

fish kills caused by the micro-organism *Pfiesteria piscicida*. Even so, state officials refuse to limit the size and number of hog and poultry confinement operations because that would mean unfair competition with other states.

- The Environmental Protection Agency in August 1997 imposed the largest fine ever on Smithfield Foods in Virginia – $12.6 million – for polluting ground and surface water. This federal action to enforce the Clean Water Act was opposed by the state governor.
- The Agricultural Export Enhancement Act allows the US government to give millions of taxpayers' dollars to help agribusiness multinationals gain a competitive edge in the world market. Using the GATT for leverage and the WTO as its enforcer, the US government claims other countries' actions are illegal 'technical' trade barriers and protectionism when these countries refuse to accept meat and milk from growth-hormone treated cattle or unlabelled genetically engineered (GE) soya beans and other genetically altered agricultural commodities.
- Countries refusing to accept GE seeds from US multinationals are coerced by threats of trade restrictions and import tariffs. 'Dumping' of surplus produce like powdered milk and chicken legs on poor countries like Jamaica undercuts and bankrupts local farmers.
- Imports of meat and cereals are likely to rise in countries that were once self-sufficient (especially China and India), benefiting exporting nations like the US, Argentina, and Australia, but making the cost of feeding poor families in developing nations higher.[1]
- US Secretary of Agriculture Dan Glickman is promoting GE crops and biotechnology as 'our greatest hope of feeding a growing world population in a sustainable way' but the right of US consumers to make informed food choices is being denied by the US government's refusal to label GE foods and food ingredients. The US government tried to overrule the National Organics Standards Board by pushing to have municipal sewage sludge, livestock confinement systems (intensive livestock units), food irradiation, and GE seeds and other products included for consideration under the proposed federal organic farming and food standards. Some 280,000 letters of protest squashed these initiatives.
- In the biggest food recall ever in the US, over 25 million pounds (1,100 tonnes) of hamburger (ground) beef was taken off the market in August 1997 because of *E. coli* 0157:H7 contamination. According to the General Accounting Office, some 9000 people die each year in the US from food-borne 'plagues'. *Campylobacter*

> is implicated as a more serious source of food poisoning from poultry than even *Salmonella*, which is found in one of every three to five birds slaughtered. One US government response is to approve the irradiation of meat and other animal produce.

Despite official complacency, US consumers, family farmers, public interest and sustainable agriculture organizations are working together and becoming a force of influence nationally and internationally. The vision of a socially just, ecologically sound, humane and sustainable agriculture is becoming a reality,[2] as local community supported agriculture and market co-ops take root in various parts of the USA.

At this crossroads for US agriculture, one path leads to the export of intensive systems and related inputs (such as antibiotics, high protein and energy feeds), which continue the non-sustainable industrial paradigm.[3] The other path leads to domestic consumption from more local, sustainable, organic and less inhumane production methods.

INDUSTRIAL AGRICULTURE'S WAY AHEAD

US agribusiness advocates, like Dennis T Avery, call industrial agriculture 'high yield' agriculture.[4] (See Chapter 2, Avery.) Its proponents, who also see biotechnology as a way to further enhance agricultural productivity and to save wildlife species and biodiversity, either deny the hidden costs or accept them. They claim the benefits – more food (and agribusiness profits for a 'hungry world') – far outweigh such costs.

They argue that human population increase means the risks and costs of intensive 'high yield' agriculture are justified (or insignificant). There is no alternative, such as organic farming, according to Avery, because it is so low yield that it will mean global famine if more wildlife habitat is not taken over to make up for the deficit per acre. Thus, organic farming is presented as a major threat to conservation and biodiversity and to the human good.

People who live by such claims structure reality in such a way that they do not know when they are lying to themselves or deceiving others. The absurdity of this new agribusiness myth was well documented in a report by the Henry A Wallace Institute for Alternative Agriculture.[5]

This report details how, in the US especially, chemically-based, intensive crop production (particularly questionable as a livestock feed-source, as José Lutzenburger and Melissa Halloway argue in

Chapter 1) harms both terrestrial and aquatic ecosystems. It also confirms that a range of alternatives to the chemically-based production model can achieve equivalent or higher yields per unit area of land with less harmful consequences.

Avery, however, suggests that industrial agriculture will not only alleviate world hunger but also help reduce population growth because people who have a better income and can afford more meat and other animal produce have fewer children.

This is an overly simplistic correlation. Smaller affluent families are, per capita, as much, if not more, of a drain on the environmental economy and energy budget as poorer families who eat little or no meat and who sustain themselves on a low-input, labour-intensive agriculture.

It is education and access to family planning programmes and the development of local self-sufficiency and sustainable enterprises, especially agricultural, not agribusiness 'high yield' farming, that will help control human population growth and world hunger.

But Avery is right in stating, 'Higher crop yields add wealth, which itself encourages lower birth rates. They also permit a shift to urban jobs and urban birth rates are almost always much lower than rural ones.'[6] The key questions, however, are which crops, how are yields increased, who grows them, who owns the land, and who controls the market.

Process not just product

The use of genetic engineering in livestock and crop production and food processing raises yet more questions and concerns, especially since the US government has essentially deregulated this new industry to give US-based multinationals a competitive edge in the world market.

Through the WTO and the Codex Alimentarius, which is drafting international agreements on food quality and safety, the prevailing values and practices of industrial agriculture may well become the global norm. US agribusiness corporations, facing international competition, will understandably resist environmental and farm animal protection legislation so long as it is illegal under WTO rules for the US to protect its own farmers from imports from other countries that have inadequate or no environmental and animal protection legislation. Without international harmonization of sound environmental and farm animal protection laws and regulations, international agreements and standards for food quality and safety are ethically unacceptable.

Unless food is labelled with country of origin and method of production (eg organic, free range, or genetically engineered), consumers will have no choice in the market place and no opportunity to support either local farmers or particular farming methods. The US government, however, under pressure from agribusiness, is resisting attempts by public interest organizations to uphold the consumers' right to know via appropriate food labelling. The forthcoming national organic food label will probably set a lower standard than many US organic farmers have achieved. This will result in unfair competition and actually mislead consumers.

US agribusiness should focus more on process, not productivity, which is the end-point of an extremely complex, biodynamic process. This process does not fit within the narrow paradigm of conventional agricultural economists.

A process focus means paying attention to the economic and health benefits of maintaining a *living soil*, which is the primary resource (coupled with pure water, improved air quality and normal solar radiation) of agriculture.[7] Agribusiness can make money helping restore and maintain soil, air and water quality, as well as the quality of livestock and seedstock (without having to resort to genetic engineering). A science, economy and ethics of remedial agricultural inputs that lead to healthier soils, crops, livestock and food should be on the corporate agenda and the mission of land-grant colleges of agriculture, 'food science' and veterinary medicine around the world.

Similarly, human medicine should establish closer links, via nutrition, with remedial innovations in agriculture and in consumer eating habits. It is absurd that the pharmaceutical and medical industries should continue to profit by selling many products and treatments that would not be needed if our soils were healthy, our food was safe and nutritious, and our diets and lifestyles tempered by the science and philosophy of biological realism and bioethics.

HUMANE SUSTAINABLE AGRICULTURE: BIOETHICAL PRINCIPLES AND CRITERIA

I believe many agribusiness groups who oppose the humane sustainable agriculture movement will support it when there is a clearer understanding of the bioethics and profitability of farming non-violently which looks beyond the short-term goals and imperatives of the world marketplace.[8] Using bioethics to evaluate developments and current practices in agriculture will facilitate the adoption of humane practices.

In bioethics, there are seven golden rules:

1 Reverential respect
2 Ahimsa (avoid harm/injury)
3 Compassion
4 Social justice and trans-species democracy (equal and fair consideration)
5 Eco-justice/environmental ethics
6 Protect and enhance biocultural diversity
7 Sustainability.

The core principle is *ahimsa*. It means non-harmful, non-injurious and non-violent action. The future of agriculture, if it is to be sustainable, must be guided by this compassionate ethic of *ahimsa*: of avoiding harm to other living beings, human and non-human; plant and animal, wild and domesticated, either directly, or indirectly by damaging the environment. All new agricultural products, processes and policies should be subject to rigorous bioethical evaluation prior to approval and adoption.

First and foremost, we need a *soil-based* agriculture that uses various crops, forages and animal species sustainably within the limits of available, renewable, *local* natural resources; and that either enhances or causes no net loss of *natural biodiversity*.

Maintaining soil and water quality and biodiversity are the basic bioethical criteria for social acceptance of those farming systems that function profitably and sustainably with or without animals. But until national and international bioethical accord and global harmonization of organic and humane farming standards and practices are achieved, great effort will be needed to protect humane and sustainable agricultural systems and communities from unfair competition and possible annihilation by industrial agriculture

Farming with less harm, consuming with conscience

Rethinking the role of meat lies at the heart of biological realism and bioethics. No other society, past or present, raises and kills so many animals just for their meat. Nor has one adopted such intensive systems of animal production and non-renewable resource-dependent farming practices. These have evolved to make meat a dietary staple, and to meet the public expectation and demand for a 'cheap' and plentiful supply of meat. This meat-producing agriculture depends on costly non-renewable natural resources and precious farmland to raise the feed[9] for these animals to convert into flesh; land that critics now believe should instead be used more economically to feed people directly. To a hungry world, such conspicuous consumption is a poor model to emulate.

Many people still believe that factory farms and feedlots help America lead the world in producing meat at the lowest cost, and that to abolish them would hurt the poor who could not afford more humanely- and ecologically-raised, organically-certified meat and poultry. A broader bioethical perspective shows that factory farms and feedlots are neither efficient nor sustainable ways of producing food for human consumption and indirectly contribute to world hunger.

Supporters of intensive animal factory farming claim that humane reforms would increase costs and put an unfair burden on the poor. Critics of factory farming are also accused of being more concerned about animals than people and against progress. Both these criticisms are wrong.

Dumping: the real costs

The real costs of factory farming range from price supports and subsidies at taxpayers' expense to the demise of family farms and rural communities, from a waste of natural resources to public health risks and costs, and farm animal stress, disease[10] and suffering. When added to corporate monopoly, these hidden costs have aggravated rather than alleviated poverty and malnutrition nationally and internationally. Agribusiness does not pay the real costs of factory farming and its 'high productivity' is neither efficient nor socially or ethically acceptable

In some countries, such as Brazil, livestock production is a major hedge against inflation. Overproduction cycles, however, depress world market prices, fuel deforestation and other forms of environmental degradation. Price supports and subsidies to producers, especially in the developed world, encourage overproduction and cause further distortions and inequities in world market prices. One serious consequence is the 'dumping' of meat, dairy and other agricultural products in other countries. These are sold to processors and wholesalers at prices much lower than local farmers can get for their own similar produce.

Legally, 'dumping' is defined as putting products on the market for sale at a price below the actual costs of production. This definition should be broadened to include all marketing activities that undermine regional self-sufficiency, national sovereignty and local sustainable productivity of the same or similar commodities and services. The fair market price of agricultural commodities and services should reflect all costs, including social and environmental. More equitable trade policies could then be established based on *full cost accountability*, and markets encouraged or protected as appropriate.

Raising tariffs and other forms of 'protectionism' by any country to protect its own farmers, however, is an illegal 'technical barrier' to trade under WTO rules. Nonetheless, local farmers raising food and feed for domestic consumption should have their market protected and fair market prices guaranteed, provided their farming methods are humane, socio-economically just and ecologically sound and sustainable.

Industrial-scale, intensive systems of livestock and poultry production are being promoted in 'developing' countries as means to increase agricultural production and 'efficiency'. It makes no long-term sense except to those who manufacture and profit from selling all the 'inputs', from drugs and vaccines to feed and equipment that these animal factories depend upon.

Bioethical trading

Bioethics considers social and environmental as well as economic factors in production. It leads to incentives to promote the most ecologically appropriate farming methods and choice of crops for domestic use and for export. Policies that mean one country or region harms its constituents or its ecology and natural resources by investing in large-scale production of grain, livestock, cotton or some other commodity, and then compounds this harm by 'dumping' such produce on the world market and lowering the fair market price, are unacceptable.

Socio-economic, environmental and ethical concerns cannot be ignored by the WTO and in future GATT negotiations. To farm with less harm clearly has international ramifications for equity and world trade. All agricultural development needs to be considered from an ethical as well as an economic perspective, and constraints applied for the good of all. The same is true for new GE products, such as analogue cocoa, vanilla and nut oils. The production of these will harm those countries dependent on raising these products naturally for export revenues, needed in part to pay off the interest accrued by too often misguided development loans.

Moreover, unless the WTO protects and encourages local agricultural self-sufficiency in poorer countries the world market will become increasingly dysfunctional and may well collapse if poverty and socio-economic inequities and strife continue to spread under the compounding pressures of population increase and environmental degradation. Applying bioethics could help them do so and every nation and region maximize productivity and minimize adverse environmental and socio-economic consequences primarily by encouraging mixed farming systems (including agroforestry and aquaculture) that are most appropriate ecologically and culturally for each biogeographic region.

FARM ANIMAL FEEDS AND WASTES

Meat industry defenders deny that feed imports from developing countries contribute to hunger and poverty. They insist that many feed imports are crop by-products[11] of cash crops grown for export, such as sugar cane, molasses, palm kernel cake, cotton oil seed cake, soya bean cake, rice and wheat bran, and rice polishings. In reality, this market for by-products simply perpetuates unsound agricultural practices in poorer countries, undermines traditional sustainable farming systems and uses up good land that should be used to feed people first.

Imported feed enables farmers to rear far more animals than the land can sustain from local resources alone, and is a major support-structure of intensive livestock and poultry production. It is ethically, economically and ecologically unacceptable, partly because of the animal waste produced. This should, but is not and cannot, be returned to enrich the land in other regions and countries from which the animal feed originated. Such animal waste has become a costly environmental management hazard and is a key indicator of bad farming practices and agricultural policy. Nitrates, phosphates, bacteria, antibiotic and other drug and feed additive residues, such as copper, arsenic and selenium in farm animal excrement, overload and pollute the environment and food chain. A recent US government report indicates livestock produce 130 times more waste than the entire human population in the US.

Another problem is the enormous volume of waste that the meat industry calls 'animal tankage'. This is the dried and processed residue produced by rendering plants from the remains of dead, dying, diseased and debilitated livestock and poultry, condemned and unusable body parts and even the remains of road-kills and cats and dogs from animal shelters. In the US alone, it totals around 44 billion pounds annually (nearly 200,000 tonnes). Slow, low-heat rendering neither sanitizes nor rids animal tankage of potentially harmful organisms, heavy metals and other hazardous residues. Farm animals, companion animals and consumers are all put at risk since this by-product of animal agriculture is added to pet foods, livestock and poultry feeds (which may also include animal manure). It is even sold as fertilizer for farm, home and kitchen gardens. Studies have linked bacterial food poisoning in humans with this industry practice of including animal tankage by-products and poultry manure in farm animals' food. The tragedy of 'mad cow disease' in the UK is principally a product of an increasingly dysfunctional food industry. Obviously, if consumers eliminated or reduced their consumption of meat and other animal produce, the magnitude of these problems would be significantly reduced with great economic savings, because there would be fewer animals being raised.

Bioethical solutions

Those who believe that farm animals do not play a vital ecological and economic role in sustainable crop production and range management (See Chapter 16, Gold) are as wrong as those who claim that intensive livestock and poultry production is bioethically acceptable because it causes no harm. It is the industrial factory-scale system that the animal component of modern agriculture has evolved that is bioethically unacceptable.

The rightness or wrongness of meat eating or killing animals is not the central issue or primary bioethical concern. The primary concern is to implement less harmful alternatives to contemporary animal agriculture in the US and in other countries that have factory farms and feedlots. The solution lies in the adoption and public support of less harmful organic and other alternative, sustainable crop and livestock production practices that are humane and ecologically sound – what Lutzenburger calls regenerative agriculture.

A reverential respect for life and for the land is the guiding bioethical principle of a humane, socially just and sustainable agriculture and society. Questioning agricultural practices, including new developments in genetic engineering biotechnology that may cause harm, is not unscientific or designed to obstruct progress. The essence of progress is to apply science and the seven golden rules of bioethics to new developments so as to cause the least harm and the greatest good to the entire life community of Earth. We cannot sacrifice the good of the environment or of rural communities for short-term corporate profits. If we do, society will suffer, if not this generation, then the next. Likewise, we cannot sacrifice the good of farm animals or of the soil in the name of productivity and labour-substituting technological innovation and marketing, without ultimately harming the economy and the health of the populace.

More humane and healthier diets

We humans are a highly adaptable primate species. One feature of this is our physiological capacity to be omnivores. This flexibility in our capacity to use a wide range of food sources is universal but with cultural nuances that may have a genetic basis. For most peoples around the world, a primarily plant-based diet, with animal products as supplements or condiments, has been shown to be the keystone for a healthy life, economy and environment (see Chapter 11, Cannon). With rare exceptions, most peoples can eat and digest almost anything that other mammalian species can assimilate (except cellulose). People have developed remarkable ways to preserve and enhance the nutritive value and palatability of a diversity of natural foods. Cultural and ethnic differences in cuisine reflect biogeographic and seasonal variations in food types and availability. This ethnic diversity provides a rich cornucopia of culinary delights and is a source of new crops and food products for an increasingly cosmopolitan marketplace.

An animal-based agriculture and a meat-based diet are neither good for the planet nor for one's health. These views, now being more widely accepted and promoted by health experts in the US, confirm the connections between a healthful diet and humane and sustainable agriculture. This should do much to encourage traditional and innovative organic farming practices and ethnic foods, and help prevent the loss of biocultural diversity in world agriculture as well as in the kitchen, which is under siege by the promoters of meat and other animal produce as dietary staples.

Overproduction and overconsumption of food go hand in hand.[12] The relatively low market price of food in the US compared to other countries, coupled with increasingly sedentary lifestyles, has led to the alarming finding that one in three American adults is now seriously overweight and the average body weight is increasing.[13] That means 58 million people have an increased risk from heart disease, diabetes, cancer and other chronic ailments. Ironically, overeating has spawned a weight-loss industry that reaps $40 billion per year from American consumers, more than most countries spend on food.[14] Furthermore, the US food industry spends some $36 billion a year on advertising.

The consumer trends in industrial society toward nutritional illiteracy, agricultural amnesia, and culinary catatonia, fostered by the microwaveable frozen meal industry with its prepared and processed convenience foods and relatively meaningless ingredient and daily recommended allowances labelling, are symptomatic of the disintegration of agriculture and culture. So are the diseases of an overconsumptive and malconsumptive (and malcontent) society that justifies health spas, costly coronary bypasses, and liposuction to remove excess calories, while the rest of the human population that might well aspire to live this way suffers from malnutrition and even starvation due in part to the insatiable appetites of the industrial world.

The first most important step to counter this for every caring person to take is to choose a humane diet and eat with conscience. This bioethical imperative is a vital component in:

- the prevention of animal suffering;
- reducing public health costs;
- restoring rural communities; and
- the preservation and restoration of farm lands and natural resources.

Encouraging people to make this caring and enlightened choice is politically controversial. It is still seen as an economic threat by those who have a vested interest in stopping real progress in agriculture.

Eighteen US agribusiness groups[15] purportedly representing the best interests of farmers, ranchers and farm animals, have publicly attacked the Humane Society of the United States' (HSUS) 'Choosing A Humane Diet' project. They claim that, 'This campaign for the first time really places the HSUS squarely in the lead of animal rights groups seeking a vegetarian society by using emotionalism to induce the public to both reduce and replace animal products with other foods.' By so doing, these groups seek to discredit the legitimacy of our concerns and to protect their vested interests in maintaining the status quo of factory farms and feedlots. We are also cast as the enemy of farmers and ranchers. Yet the real enemy is agribusiness, which has contributed to the loss of over 425,000 family farms over the past decade as animal factories and feedlots have proliferated and put smaller producers out of business.

Farming with compassion

To farm compassionately and cause less harm means a reduction in the production and consumption of meat in those countries where meat is a dietary staple. It also means improving the ways animals are raised, transported and slaughtered, and *replacing* animal protein and fat with cheaper vegetable fats, oils and proteins. If the earth's population doubles from its current 5.6 billion (of whom 1.6 billion are malnourished), maintaining the status quo and public demand for meat means the loss of biodiversity and non-renewable resources, and the attendant environmental and economic risks and costs. We have a better chance to predict and prevent these if we apply bioethics in the public and corporate policy decision-making process as well as in our own lives. And the first principle of bioethics in agriculture is like the good doctor's Hippocratic aphorism: do no harm.

If it causes less harm to the ecology of a particular biogeographic region to include some animals in the farming system to make it profitable and sustainable than it would if they were excluded, then their ***humane incorporation*** is ethically acceptable. An agriculture that accepts cruel treatment of livestock and poultry is unethical and dysfunctional.

The degree of humaneness, the quality of life afforded to both farm owners, workers and farm animals, should be the cardinal indicator of sustainable profitability and social acceptability of those farming systems that have integrated animals as essential ecological components.

The major criteria for bioethical evaluation are illustrated in Figure 3.1. These interdependent criteria are interconnected and they all converge on economics and full cost accounting. These bioethical criteria include:

- safety and effectiveness;
- social justice and equity;
- farm animal well-being;
- environmental impact (including harm to wildlife, loss of ecosystems and biodiversity);
- socio-economic and cultural impact, especially harm to established sustainable practices and communities; and

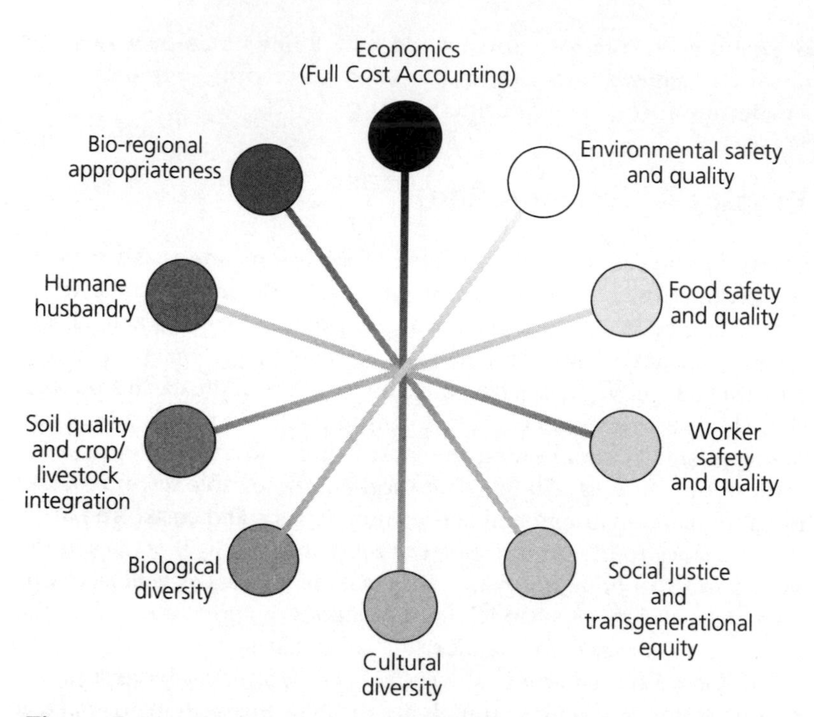

Figure 3.1 To Farm Without Harm: Bioethical Criteria and Principles of Organic Sustainable Agriculture

- accord with established organic and other humane sustainable agriculture practices, standards and production claims.

Farming with less harm and choosing a humane diet are two sides of the same coin. They will forge a strong alliance between urban consumers who care, and rural producers who share the vision of a humane and socially just agriculture and society.

Protecting nature and wildlife

'Farmers work at the junction where population, the human condition, and sparing land for Nature meet' argues agronomist Paul E Waggoner from the conservative Council for Agricultural Science and Technology.[16]

Waggoner shows how 'smart farmers' can harvest more per plot and thus spare some of today's cropland for nature – if we help them with changed diets, never-ending research, and encouraging incentives. His suggestions include:

- Calories and protein equally distributed from present cropland could give a vegetarian diet to 10 billion people.
- The global totals of sun on land, carbon dioxide in the air, fertilizer, and even water could produce far more food than 10 billion people need.
- By eating different species of crop and more or less vegetarian diets, we can change the number who can be fed from a plot.
- Recent data show that millions of people do change their diets in response to health, price, and other pressures, and that they are capable of changing their diet even further.

RESTORE THE HUMUS, RECOVER HUMANITY

The decline and demise of civilizations is almost invariably linked with the devitalization of the soil and consequent malnutrition. Today, this is compounded by variously denatured, deficient, refined, processed and adulterated foods. We human beings tend to forget that we are humus beings. From the earth we are born; to the earth we return and by the earth we are sustained. ***Humility, humanity and humus*** are words that connect and ground us in the reality of our being.

Unfortunately, the limited worldview of egotism separates us from the biological reality of our own being, and so out of ignorance we demean, neglect and abuse the earth. Caught in the delusional realm of anthropocentrism, we fail to realize that when we harm the earth, we harm ourselves. When the humus is depleted of microorganisms, becomes nutrient deficient and toxic with agrochemicals, so become our crops, farm animals and the food we consume: and so become our bodies and minds. In harming the earth we harm ourselves physically and mentally, and the worldview or mind-set responsible also harms us morally and spiritually.

When we recover our humanity and humility we rediscover the wisdom of living in harmony with the earth. Through the inter-communion of reverential symbiosis we come to understand and respect, as the laws of Nature, all the relationships and processes that maintain and sustain the life community. And we are secure in the knowledge that we are part of that which is forever being renewed, as the self is forever. Obedience to these laws enables us to participate in a creative and mutually enhancing way and by so doing develop appropriate technologies that do not cause harm to ourselves and other sentient beings, and which ideally help enhance the life, health and beauty of the earth.

Our scientific understanding of ecology and evolutionary biology provides a rational, ethical basis for what we regard spiritually as our sacred connections and shared origins, since we are part of the same Creation as all other sentient beings. This spiritual kinship leads us to acknowledge the intrinsic value and inherently divine aspect of every being. We neither rob animals of their dignity nor their sanctity and right to be themselves and fulfil their cosmic purpose.

As we humans come to see that most evil in the world comes from our ignorant self-centredness, we may, with Nature's help, mature into a Creation-centred being. We should reflect on the wisdom of Albert Einstein, who surmised that, 'The significant problems of the world cannot be solved at the same level of consciousness at which they were created.'

Our pathological anthropocentrism has pervaded our major religious and cultural institutions and caused great harm for millennia. The recovery of agriculture and civilization lies in the transformation of our consciousness, combining compassion and reason, conscience and science. This will herald a new epoch in human evolution and in the refinement and metamorphosis of the human spirit. An auspicious beginning is to respect the living soil as a primary life-giver and sustainer, and to farm and to consume accordingly, with less harm and greater care, harmony and veneration.

We will know when we are on the right path again when we farm with less harm; agrochemicals are rarely used; when livestock factories are gone, along with the notion that meat is an ethically acceptable and necessary dietary staple; and when food – its production, marketing and consumption – regains those sacramental elements of stewardship, thanksgiving and communion, and is no longer regarded simply as a profitable commodity.

NOTES

1 Consultative Group on International Agricultural Research (1997) *The World Food Situation: Recent Developments, Emerging Issues and Long-Term Prospects*, CGIAR, Washington, DC

2 See Lockeretz, W (ed) (1997) *Visions of American Agriculture* Iowa State University Press, Ames, IA; Berry, W (1978) *The Unsettling of America: Culture and Agriculture* Avon, New York; and National Research Council (1989) *Alternative Agriculture* National Academy Press,Washington, DC

3 Some mega-factory farming operations have moved across the border to Mexico where labour is cheap and environmental regulations non-existent, for example, North Carolina-based Murphy Farms which plans to set up 40 or more hog factories there
4 See Dennis T Avery (1993) *Biodiversity: Saving Species with Biotechnology*, Hudson Institute, Indianapolis
5 See T I Hewitt and K R Smith (1995) *Intensive Agriculture and Environmental Quality: Examining the Newest Agricultural Myth*, Henry A Wallace Institute for Alternative Agriculture, Greenbelt, MD
6 See Dennis T Avery (1995) *The Farmers' Plea to Environmentalists*, Hudson Institute Agricultural Conference paper
7 The hidden costs of industrial agriculture and alternatives are explored in M W Fox (1995) Second edition *Agricide: The Hidden Farm and Food Crisis that Affects Us All*, Krieger Publishing Co, Malabar, FL
8 Agricultural economist Harold F Breimyer argues in 'Farms and Farming in the Next Century: A Futuristic View' *Small Farm Today* June 1995, pp 11–12, that 'The 21st promises to be a century of biomass agriculture. Industry's voracious demand for biomass products of agriculture will have two startling effects. The first is to draw farming resources out of animal agriculture, converting us all into pasta-and-vegetable eaters. The second is to make us as a nation more protective of our farmland resource than ever before... ...Our cropland will be farmed more intensively than it ever has been in the past. Farming practices will be much more labor intensive than they now are. Our topsoil will be meticulously protected against erosion or other damage...'
9 According to the FAO, *FAO Production Yearbook 1990* (Rome 1991), some 800kg of grain is used to feed livestock in the US to meet the annual per capita average consumption of 42kg beef, 20kg pork, 44kg poultry, 283kg dairy products and 16kg of eggs
10 According to a 1986 report by the Office of Technology Assessment, animal diseases cost US agriculture $17 billion annually (*Feedstuffs* March 14, 1994)
11 Various domestic crop, food and beverage industry by-products (like citrus pulp and brewer's grains) do play an important role in providing feed for integrated livestock and poultry production
12 Meat consumption in the US continues to increase. The average American consumed 204 pounds of meat in 1993, including 65.4 pounds of beef, 52.3 pounds of pork, 68.6 pounds of chicken, and 17.8 pounds of turkey. See *Illinois Agri. News*, July 15, 1994, pC2
13 R J Kuezmarski, et al (1994) 'Increasing prevalence of overweight among US adults' *J Amer Med Assoc* 272, pp 205–11
14 NBC news report, July 19, 1994
15 These groups are as follows: American Farm Bureau Federation, American Feed Industry Association, American Meat Institute, American Sheep Industry, American Veal Association, Animal Health Institute, Animal Industry Foundation, Egg Association of America, Holstein Association of America, Livestock Marketing Association, National Broiler Council, National Cattlemen's Association, National Livestock Producers

Association, National Milk Producers Federation, National Pork Producers Council, National Turkey Federation, United Egg Association, and United Egg Producers

16 Paul E Waggoner (1994) *How Much Land Can Ten Billion People Spare for Nature?* Council for Agricultural Science and Technology, Ames, Iowa

4 Future Agriculture: Giant or Gentle?

Christine Townend[1]

Farming in the industrialized world uses what I call the 'giant' approach to agriculture. It involves a reduction of labour inputs to increase profits and the relegation of animals to the status of production units. Alongside this, there is usually concentration of capital, as small family or village farmers are replaced by agribusiness conglomerates. Agriculture becomes the province of the corporate sector, rather than an integral, cultural aspect of living. This is occurring worldwide – both in extensive agriculture, as in the Australian wool industry, where sheep are left in paddocks largely to fend for themselves, and in the intensive industries, where animals are held indoors in confined conditions, with very limited human contact.

In India, we can still see the 'gentle' approach to agriculture. This is small-scale agriculture in which the human–animal interrelation is close. The agri-'culture' is inseparable from the other events of life and inseparable from an understanding of a human's relationship to the animals and nature. We see it in the many village farming communities where cattle are primary in providing draught animal power, dung for fuel, building and fertilizer, and milk. The cattle are kept in close proximity to the humans and there is an understanding between the two kingdoms of nature. The animals are a part of the daily life and religious ceremonies of the people.

India stands at a cross-roads in her relationship to agricultural animals. The country could become one giant slaughterhouse if the WTO and commercial interests can break down her people's traditional resistance to the slaughter of animals. Now, as never before, India needs the support of the international community to avoid this and so that small-scale, human and animal-centred agriculture can survive. In India there are extremes in every facet of life – extremes of cruelty and extremes of compassion practised towards humans and animals. The human/animal interrelation is important in itself because we will never have a world at peace if we cannot break down the barriers and misunderstandings which humans presently hold towards animals.

In Indian streets, you can walk up to an unknown cow, extend your hand and let it be smelt and licked. This act of communication between two different kingdoms of nature takes place in an atmosphere of trust and mutual curiosity. In Australia, the cattle flee in terror if a human enters the paddock. You might think that two different species were involved, instead of two different cultural attitudes. However, I am hopeful about maintaining the human/animal interrelation in India because of two factors:

1. The practice of 'gentle' agriculture is a part of the cultural relationship to nature and cannot easily be separated from daily village life, even though this is under challenge as never before.
2. There is a vigorous people's movement to defend the needs and interests of India's animals.

RESPECT OR ECONOMIC IMPERATIVE?

Western anthropologists often suggest that the respect for the cow in India was merely a result of historical and economic events. For example, Marvin Harris argues that the cow had to be treated as sacred as this protected her from being butchered in times of drought, when the need was to preserve her for future breeding.[2] This is a derogatory attitude which does not give any credit for India having an advanced ethic in relation to animals and nature. The original Aryan nomads who invaded India around 1750BC did bring with them the tradition of beef-eating. But about 2500 years ago, when Buddha and Mahavira, the founders of Buddhism and Jainism, taught in India, the notion of *ahimsa* (harmlessness), which also included within its strictures the practice of vegetarianism, became widespread.[3] Historians debate about the origin of the seemingly contradictory change which happened as the priests changed their role from the takers of animal life to vegetarians who would be polluted if they killed animals. Some writers maintain that the *Rig Veda*, a sacred scripture of the early Aryans, already contained within its writings the seeds of *ahimsa*, and that Mahavira and Buddha built upon these already dormant notions.[4] In any case, says Jeremy Rifkin, 'This was, indeed, a significant gesture for a religion whose leaders descended from cattle-raiding warriors of the steppes.'[5]

It is the love and respect for the cow which caused such a close and sensitive relationship throughout the centuries, not the economic, technological, geographic and demographic physical conditions which forced cow protection to be practised. If cow protection was merely a result of economic necessity, then we would find other cultures which are totally dependent on animals for food, fibre and draught power would also have developed the concept of *ahimsa*, vegetarianism and cow protection. Yet in Christian Europe, although animals were necessary for virtually every aspect of life, a very different view of the human–animal interrelation developed in which humans were seen as the God-given masters and dominators of all other kingdoms of nature.

Writing about China, Weber argued that there were opportunities present in China for the development of capitalism, just as there had been in Europe, but simply because there was a particular political and geographical climate, this did not automatically give rise to capitalism. Weber looked at the condition of money supply, cities, the literati, agrarian policy, political and military structure, and concluded that some conditions were unsuitable, and others suitable, for the rise of capitalism. Indeed, he claimed, it was because Chinese Confucian and Taoist religion placed an emphasis on the human place in nature and its necessity to harmonize with the surrounding natural world, that capitalism did not develop at that time. In Europe, however, Protestantism led to a system of beliefs in which humans believed they were above the rest of nature, and it was their duty to work hard to dominate it.[6] It was the system of values which India developed in relationship to animals, and the natural world, which led her on a different path to that of Europe.

Critics of cow protection claim it is ridiculous that the life of a cow cannot be taken under the Indian law of most states, and at times we feel exasperated at our shelter in Jaipur when a suffering bovine cannot be destroyed. However, this is because the very high principle of honouring the lives of the cattle has become perverted and corrupted by political interests. Indeed, we should be impressed that animals are so important to Indian society that the life of a cow is a political issue. If slaughterhouses became a political issue in Australia then the ecologists and animal rights campaigners would be well on the way to winning.

Even so, there are extremes of cruelty to cattle in India. For example, in some states they are driven long distances to slaughter, and the way they are slaughtered leaves much to be desired. (See also

Chapter 10, Gandhi.) But the cultural tradition of respect is there, no matter how it fails; and it does fail, as is the case with human welfare as well. The ideal is there but the ideal is not always expressed, just as the children of India are also not treated ideally, but are much valued and loved.

AUSTRALIA'S GIANT AGRICULTURE

In Australia, the 'giant' approach to agriculture followed the trends of other Western nations. As agriculture became more and more focused on the profit motive, and less and less concerned with quality of life, humans became more and more isolated from animals. One example of this is in the Australian wool industry which involves enormous animal suffering.[7]

There has been a gradual erosion of this interrelation in the Australian wool industry as the ratio of labour per sheep unit has increased due to economic pressures. Instead of small flocks with intensive shepherding, resulting in proper veterinary care, nutrition and protection from the elements, the labour component is low as family farms are forced to acquire more and more land and animals to survive financially. The corporations use agriculture solely to make money, ignoring the traditions in which, generally, smaller flocks were closely shepherded. There are on average about 2000 sheep to every labour unit but 8–10,000 sheep per person is not uncommon.[8] If one person is in charge of 2000 sheep, it is impossible to check each sheep individually each day. If the farmer did nothing else but inspect the sheep for eight hours of the day, it would mean that he or she could only give 0.24 of a minute to each sheep. In reality, there are many other jobs to do, so the sheep are left unattended for long periods. It is cheaper and easier to lose some sheep than to pay the costs of proper supervision.

In mid 1998 in outback New South Wales, for example, the price of an old merino wether (a castrated ram) ranged between $A8–$A20 and the price of a young merino ewe with a good fleece was about $A40, but the cost of a visit from the vet for a grazier was around $A150. For this reason, sheep losses 'on the farm' in Australia are massive. Some 6.2 per cent of the national sheep flock die in the paddock each year.[9] In addition, 20 per cent of lambs die due to lack of proper care.[10,11] This is largely due to neglect – inappropriate shearing in mid-winter, failure to provide proper shelter, and brutal surgical procedures designed to cut labour costs. The worst of these is mulesing, in which slices of flesh and wool are removed around the tail area of the lamb (or sheep) to create a tight area of scar tissue, without wool or skin wrinkles. This means that there is a clean area under the tail and the grazier does not need to jet (spray dip), crutch (shear wool under tail) and check the sheep so often to prevent fly-strike. Mulesing is the most brutal and bloody response of the 'giant' approach to agriculture, where the only thing that counts is a successful commercial venture, and the life and suffering of the sheep play no part in this formula.

Seventy years ago, almost 40 per cent of Australia's population lived in rural areas, but today, 85 per cent of the population is urban.[12] In Australia today, the small, family farms are being squeezed out and crushed due to 'market forces'. The Bureau of Agricultural Economics describes the frightening trend towards 'giant' agriculture:

> *There has been a decline in the percentage of the population residing in rural areas and in the number of persons employed in agriculture. This reduction has been accompanied by a marked increase in average farm size and an increase in use of capital equipment and purchased inputs such as fuel, fertilisers and pesticides. There has been a substitution of capital and other inputs for labour.*[13]

'GIANT' THREAT TO INDIA

This trend towards 'giant' agriculture has already begun in India with trade liberalization. The journal *Poultry World*, under the heading 'Great Prospects for the Indian Poultry Industry', estimates that by AD2000 poultry consumption will have doubled to 820 million birds per annum in India. The target consumers are 'the fast growing middle class of over 250 million potential customers'.[14] Although India still has a serious commitment to respect the other kingdoms of nature, it is unlikely that most people understand or care when they see new factory farms built in the countryside. A factory farm may well be perceived as a kinder way to keep animals, due to lack of information about the method of production. However, there is still significant public resistance to the slaughter of animals and attempts to open new slaughterhouses. India still has the largest number of vegetarians by choice of any country, and has expressed legislatively her commitment to the tradition of *ahimsa*. When India became independent in

1947 the new Constitution included a ban on cow slaughter in Article 48. The principle of animal protection is incorporated in another clause, Article 51(g): 'It shall be the duty of every citizen of India to protect and improve the natural environment, including forests, lakes, rivers and wildlife and to have compassion for living creatures.'

The Animal Welfare Board of India (AWBI) was the first of its kind to be established by any government in the world in 1962 under the provisions of Section 4 of the Prevention of Cruelty to Animals Act (1960). It has 28 members including six MPs, humanitarians, representatives of the government of India, and representatives of Societies for the Prevention of Cruelty to Animals and animal welfare organizations. Its policy is vegetarianism.[15]

There has been a continuing debate between secular Hindus, Muslims and traditionalist Hindus since Independence about the slaughter of cattle.[16] Recently the residents of Narela, an area of Delhi, pleaded against the government's plan to build a slaughterhouse in their area. The judgment passed in the Court by Shri C K Chaturvedi said in part:

> *This fundamental Duty in the Constitution to have compassion for all living creatures, thus determines the legal relation between Indian Citizens and animals on Indian soil, whether small ones or large ones. This gives legal status to the view of ancient sages down the generations to cultivate a way of life to live in harmony with nature. Since animals are dumb and helpless and unable to exercise their rights, their rights have been expressed in terms of duties of citizens towards them. Their place in the Constitutional Law of the land is thus a fountainhead of total rule of law for the protection of animals and provides not only against their ill treatment, but from it also springs a right to life in harmony with human beings.*

In 1990, Prime Minister V P Singh said in a famous speech that: 'Hinduism is above all a religion of synthesis. It has united the animate with the inanimate, the soul of one with the soul of all, the *atma* (spirit) with the *paratma* (God).'[17]

It is hard to imagine a judge in an Australian court giving the rights of animals priority over the rights of the government to establish a slaughterhouse. And if an Australian Prime Minister started to talk about the soul of one being the soul of all, he would probably be dismissed from office as being mentally unstable. Unfortunately, Western materialism has dulled our senses to such a degree that we see such attitudes as being backward and reactionary, instead of acknowledging that in many ways this philosophy provides an inspiring example to be followed.

'Gentle' Economic Benefits

The 'gentle' approach to agriculture not only benefits humans and animals, but is also economically important to India. Three quarters of India's population is rural. In 1990–91, the number of marginal, small and medium scale holdings of 10 hectares or less was 103,611,000, but there are only 1,668,000 large-scale holdings over ten hectares.[18]

Many of the villages are not connected by road, and draught animal power is the only means of taking a sick person to hospital. The bullock, buffalo or camel cart is usually viewed as something backward and cruel. In reality, despite all its shortcomings and given the many improvements which are needed, the bullock cart in traditional Indian village life represents the essence of 'gentle' agriculture and the human/animal interrelation at its very best in terms of ethics, economics and sustainability.

The organisation CARTMAN (Centre for Action, Research and Technology for Man, Animal and Nature), based in Bangalore, under the guidance of Professor N S Ramaswamy, has campaigned for two decades to promote small-scale agriculture using appropriate technology – namely, the modernized bullock cart. CARTMAN argues that there are about 80 million draught animals in India. One bullock can provide the power equivalent of 0.5 hp. Thus draught animals make available 40 million hp, ploughing 67 per cent of cultivated land and hauling in a year freight equivalent to 25 billion journeys of 1 kilometre length and 1 tonne weight. If this work were to be carried out by mechanized power, the annual requirement of petroleum products would be six million tonnes. Thus, there is a notional annual saving on petroleum products amounting to Rs. 9000 crores ($US2.3billion; 1 crore = 10 million).[19]

The most ecological and sustainable system is in the traditional Indian village where the human/animal interrelation is strong. The cow provides milk, and the bullock provides draught animal power. The dung is used for fuel, fertilizer and building material. The calf is not taken from the mother and killed at birth. Indeed, traditional Indian breeds of cattle have evolved so that they do not even let down the milk unless it is first sucked by the calf. The unproductive animals were not slaughtered but were kept in *pinjrapoles* and *gaushalas* (cow shelters), fodder being provided by those who wished to donate towards their upkeep.

However, as modern capitalist economics intervene, this model is rapidly being destroyed. If we directed resources into developing 'gentle' agriculture, instead of squandering the resources on the development of ugliness, we could upgrade and improve Indian agriculture so that it was more productive, whilst at the same time preserving its human and animal face.

For example, the estimated value of dung produced in India each year from cattle is about Rs.10,000 crores. 'The replacement of animal and manual energy would entail an additional investment of 25 to 40 billion dollars on electricity generation. Over and above this, it would entail loss to the farm economy of organic manure, and cheap fuel'.[20] Draught animals are recognized by the UN as one of the 14 sources of renewable energy and there are programmes in India to use dung more effectively. If dung could be used to provide biogas, huge energy savings could be made. Biogas not only provides fuel but also transforms agricultural waste into concentrated organic matter. This is a much more efficient fertilizer than raw dung. About two million small biogas plants could save about 5.7 million tonnes of wood being burnt each year, valued at Rs 285 crores per annum. China has reportedly set up over eight million biogas plants but India still has a long way to go.[21]

Animals in cities

Five years ago, the drive from Delhi to Jaipur presented a vista of small villages, bullocks ploughing the fields and camel carts gliding along the road. Now, much of the countryside is obscured by huge ugly factories built by multinational corporations as a policy of trade liberalization is pursued by the Indian Government.

For some of India's large cities it is already too late. The 'giant' concept has already replaced the gentle. To make cities and towns as we, the people and animals, want them to be, we would have to ensure that the camel carts, bullock carts, flocks of goats, packs of dogs, and cows, pigeons and monkeys remain, despite the nuisance they sometimes cause. We could have 'animal zones' in cities, just as we have parks and malls. In the animal zones, there could be enclosures for cattle, and for horses. Some streets and some lanes of highways should be set aside for the draught animals. Since India is just beginning to construct a system of highways, it is an ideal time for planning the inclusion of animal lanes. In these, the villagers could safely travel without fear of being killed by mad truck drivers on drugs, or a new middle class driving Hondas and Audis.

Having animals on the roads in their dedicated lanes might even reduce traffic accidents thanks to alleviation of stress, which can be dispelled by the aura of peace and serenity which surrounds these great and beautiful creatures. I know that if I go and stand in the cattle shed at our shelter all my anxieties will dissolve in the company of those animals, their large black eyes, long ears, softly chewing the cud, gazing placidly from some other world, some other level of consciousness.

India could become a tourist Mecca if she really promoted and maintained a small-scale, ecological agriculture. Tourists do not want to see clone cities clogged with traffic and high-rise buildings. They want to see the world as it should be, the grace and beauty of a bullock and his master working in perfect harmony and understanding together as they till a difficult corner of a field. I do not mean that exhausting human and animal labour should be preserved for the pleasure of tourists, but that working with nature is a pleasurable experience if the time-span of the labour is not extended beyond the capabilities of the labourers. Indian agriculture needs the development, in the field, of many appropriate technologies to make it more efficient and comfortable, without destroying the sustainability or traditions. Improved bullock carts, for example, would increase the comfort of the animals and save costs for the farmers.

To maintain and improve small-scale, ecological, ethical agriculture in India also requires many reforms. Land reform is one of the most important. It is now commonplace in India that as the value of the rural land increases, due to development pressures, the legal restrictions which prevented agricultural land from being alienated are also being removed. The Research Foundation for Science, Technology and Ecology, campaigning in India for human-centred development, instead of development based on commercial forces, points out:

> *One example of countries being forced to destroy the environment is by trade liberalisation measures in the undoing of land reforms and the removal of land ceilings which is destroying the ecologically managed small farms, and converting agricultural land into non-sustainable large export oriented farms. The States of Maharashtra, Karnataka, Madhya Pradesh, Gujarat have changed their land legislation since the new*

> *economic policies were introduced in India in 1991 to make conversion of land from agriculture to industry easier, and to remove ownership limits on land for export production.*[22]

MAINTAINING THE 'GENTLE' APPROACH

We need to maintain, or return to, a 'gentle' approach to agriculture in which the human/animal interrelation is paramount. In this, the 'giant' approach of one partner (the human) exploiting the other is replaced by an equal partnership in which animal and human each learn from the other. In primeval times, the relationship was balanced because human and non-human animals feared each other, and because each was subject to the same forces of nature, each species dwindling or multiplying according to the seasons and the success of evolution. But in the new relationship there will be a balance, not because of the intervention of nature, but because of the intervention of love, respect, and humility. Each equal partner will love and learn from the other. There will at last come a time when cross-species communication becomes possible in a way we have never before imagined. This will happen when each partner in the relationship gives time to the other. Time is the all-important gift of any relationship. The animals already give us time, but we do not give it to them. When we too give it, then we can start to learn from them.

As the new relationship becomes widespread, new centres may be developed where the study of animal powers becomes a serious discipline. For example, we now know that keeping a pet reduces the incidence of heart attack and high blood pressure, and that severely disabled people are helped to recover or to live more normally through a close relationship with animals. Already many experiments are underway to determine the nature of animal consciousness. Through scientific attempts to communicate with animals we will learn from them much about the many layers and levels of consciousness, not all of which are readily accessible to the human mind. We will also find the new and intimate relationship with animals leads to many other material benefits. In his book *Seven Experiments that Could Change the World* (Riverhead Books, 1996), biologist Rupert Sheldrake states that scientists have still not solved the riddle of how dogs know when their owners are coming home, or how pigeons find their way back to base. He postulates that if time was spent in finding the answers to these mysteries, among others, the world could be changed.

We are fighting now for the very right to live as we want to live in the company of animals. We must restore the human/animal interrelation in Western nations, and where it still exists, as in India, we must not let it be lost. If, in our humility, we can be equal partners with the other creatures of this world, if we can share our lives more fully and more generously with them, then we ourselves will be enriched both spiritually and materially.

This is where India's relationship to the animal kingdom is of interest to us as a positive light on the hill, and one day, surely, the gentle approach to agriculture will replace that of the giant. The slaughterhouses will be turned into gardens in memory of the animals we needlessly killed, and we will wonder at the brutality of the past.

Notes

1 My thanks to The Australian and New Zealand Federation of Animal Societies Inc., PO Box 1023, Collingwood, VIC 3066, and Animal Liberation SA and NSW who provided collated information on Australian agricultural statistics and Codes of Practice; and to CARTMAN, Bangalore, for information on the bullock cart as appropriate technology

2 Harris, Marvin (1974) *Cows, Pigs, Wars and Witches*, Vintage Random House, New York, 1974

3 Tahtinen, Unto (1983) *Ahimsa: Non-violence in the Indian Tradition*, Navajivan Publishing, Ahmedabad

4 Srinivas, M N (1968) *Social Change in Modern India*, University of California Press, Los Angeles, p 28

5 Rifkin, Jeremy (1992) *Beyond Beef*, Viking, Australia

6 Weber, Max (1964) *The Religion of China*, Macmillan, New York

7 Townend, Christine (1985) *Pulling the Wool*, Hale and Iremonger, Sydney

8 Nankivell, Paul, 'Australian Agribusiness: Structure, Ownership and Control', *Journal of Australian Political Economy*, No 5, July 1979

9 Australian and New Zealand Federation of Animal Societies (1989) 'Extensive Sheep Industry', *Submission to the Senate Select Committee on Animal Welfare*

10 Livestock and Grain Producers' Association (1978) *Sheep Production Guide*, NSW, p 72

11 Miller, B (1991) 'Pregnancy and Lambing', in D Cottle (ed), *Australian Sheep and Wool Handbook*, Inkata Press, Melbourne

12 Taylor, J A (1996) *Animal Production in Australia*, University of Queensland, July

13 Bureau of Agricultural Economics (1983) *Rural Industry in Australia*, Canberra

14 Taken from a forthcoming book by Mark Gold; *Poultry World*, Vol 12, no 7, 1996

15 Chairman, Lt Gen A K Chatterjee (Retd), in his speech to 'National seminar on Animal Welfare' held in New Delhi on 25–28 November 1997
16 Simoons, Frederick J (1980) 'The Sacred Cow and the Constitution of India', Robson, J R K, *Food, Ecology and Nature*, Gordon & Breach, Science Publishers, New York, p 120
17 Cragg, Claudia (1997) *The New Maharajas*, Rupa, p 11
18 Thukral, Dr R K (ed) (1998) *Jagran's India At A Glance*, Dainik Jagran, New Delhi
19 CARTMAN, *Modernisation of Animal Drawn Carts*, CARTMAN, 17 E Main, Koramangala VI Block, Bangalore 560 095
20 Sootha, Dr G D, 'Highlighting Objectives', in *National Conference on Draught Animal Power Systems*, CARTMAN, Special Issue, June 1994
21 Ramaswamy, N S *Livestock Agriculture and Sustainable Development*, CARTMAN, ibid
22 Shiva, Dr Vandana, Jafri, Afsar H, and Bedi, Gitanjali (1997) *Ecological Cost of Economic Globalisation*, Research Foundation for Science, Technology and Ecology, New Delhi, p 4

5 Barons of the New Millennium

John Vidal

The Vale of Pewsey in Wiltshire is one of the longest-inhabited places in Britain. When, almost 200 years ago, the author William Cobbett looked on it from the rolling downs, he called it his promised land. Cobbett was a passionate believer in small-scale farming and in the value of communities.

In the spring of 1998, I walked a ten-mile transect across his promised land to see what had happened to it. On the hilltops, I found a nobleman who owned and had turned to wheat 4000 hectares of downland. I estimated he was getting around a million pounds a year from EU subsidies. His sideline was pheasant shooting and racing. His friends were coming from all over the world. He bought his inputs from South America or wherever the world market was most favourable and his produce was going right the way round the world in return. Cobbett would have recognized nothing.

The nobleman was proud of what he had achieved. He felt he had restored these downs. In the main, just five people worked his thousands of acres. Three of them were tractor drivers in their 60s and they drove with satellite navigation. In the cabs of these monstrous machines they had absolutely no sense of the world outside. These old men, who had worked their life out for a succession of farmers, were the remnants of our preindustrialized farming. They were intensely loyal to their employer, but speechless about what had happened to the Vale of Pewsey.

Further south on my walk I met a small farmer with about 200 acres who was getting out of the industry. He could not take any more. He had 20 cows, a few acres of vegetables and a few sheep. It was mixed, traditional farming and he could not make it pay. Why was he leaving, I asked? Because, he said, the only way he could survive would be to have 100 cows but how could any man know the names of more than 30 cows. His farm land would be divided up by the big farmers or become little pony paddocks for the newcomers with a few hundred thousand pounds to spend turning old farmworkers' cottages into weekend homes.

After a long walk through a deserted landscape, I ended up in the Marlborough McDonald's, which seemed fitting. McDonald's, and other great enterprises that provide us with food, need intensive farming and the 'cheap food' that only the highly capitalized and subsidized landowners can provide. It needs 100 million eggs a year, hundreds of thousands of tonnes of beef and many million chickens. And it wants them as cheaply and regularly as possible.

McDonald's was recently engaged in England's longest civil trial. The company sued two people who dared to challenge how it operated, where it got its resources, and how its suppliers treated the land, their workers and their environment.

It had fallen to me to write a book about the trial, which meant reading 40,000 pages of court evidence. The animal welfare sections shook me. McDonald's insisted that their suppliers engaged in the best practice for the industry, yet much of the evidence was of animals suffering terrible stress, of widespread disease, abnormalities and misery.

So best practice was cruelty and disease on a great scale. It was sickening. What made it worse was listening to the industry justifying its practices. Yes, said one McDonald's executives, 'as a result of the meat industry the suffering of animals is inevitable'. The judge found many of the practices unacceptable and cruel but at the time of writing – more than a year after the verdicts – McDonald's has not, to my knowledge, changed its practices globally. In the UK, however, it is already in the process of altering some of its requirements from suppliers to meet higher welfare standards.

So the industry (remember, McDonald's is the best in the field) knows that it is inflicting suffering on animals. Yet it can, quite legally, and despite court rulings, continue.

For animal welfare, read any other area of our industrialized food production. The British food system has become a way of systematically abusing nature, people, communities and animals. It is a scandal. My only hope is that other generations will come to see it as that, and will outlaw it completely.

Three or four things seem to be occurring at the same time. Together they are revolutionizing how we live and how we understand the world. First of all, massive intensification of production is taking place everywhere. It is not just in farming, but in every area of life. What has happened to farming is happening to salt production, wheat, cars, oil, cameras, widgets. New global centres are emerging for every resource and product. Four or five mega-companies in each continent are emerging at the expense of everyone else. How can a small hog breeder compete with the biggest pig farm in the world, now being built in the US? It will be almost the size of the city of Chicago.

You can see it in your local supermarket. Much of the food now comes from Africa, the Far East, Israel, mainland Europe or America. We now import food from 70 countries. It does not mean there is no British food, it is just that British food is being processed and is going somewhere else. Back in the Vale of Pewsey, where they have grown potatoes for centuries and still do, I found a wonderful village shop. I asked if they had any local potatoes. Sorry, they said, only from Zimbabwe. What, I asked, happened to Vale of Pewsey potatoes? Ah, said the shopkeeper, they go to Spain and then they come back as crisps.

This is what we used to call a third world economy, where resources are ripped off from a poor country and value is added by those with capital. It is shorthand for exploitation of people and place. It is happening on a massive scale now in India, China and Africa, where the global market is impacting massively on age-old peasant farming practices. Just as the English peasant in the Vale of Pewsey was driven off the land a century ago, so it is happening everywhere else.

Leading the way are fewer and fewer transnational corporations and I include here the supermarkets, who are now utterly dominating the supply, the sale and the distribution of food. The big five in Britain sell more than 60 per cent of all we eat. Their power over farmers, growers and the public is well known to be immense. What is less known is how they are stacking the regulatory bodies, setting the rules, hijacking the science, driving international and national legislation and standards in their favour, snaffling the subsidies, directing the research and using the law to counter anyone who stands in their way. They have access to government and the support of government.

How can this be? A corporation, by definition, has no morality or social sense and no wider responsibility than to make money. It is accountable to a few institutional shareholders who meet, at most, once a year and who do not give a fig as long as their money is earning. What no human can do, a corporation can. It can turn one blind eye to suffering, another to natural justice. It cannot see common wealth, or culture, or tradition. It cannot see beauty or any value in the small, the fragile or the weak.

We must be prepared to make a stand. Unless brave people go out into the fields, unless people buttonhole MPs and take on the corporations and the system that they have foisted on us, nothing will change. The only way governments are going to listen, I am absolutely certain, is when people really stand up and start shouting.

Part III
Animal Welfare

Animal welfare has become a major political issue in the EU, but it is important more widely in Europe and, perhaps uniquely in Asia, in India. In Chapters 6–10 the practical politics, concrete agendas for animal welfare and larger effects of factory farming are laid out.

6 Why I Do Not Eat Meat

Chris Mullin

I gave up eating meat some time ago. Not because I dislike it or on health grounds but because I could not justify the appalling conditions in which most farm animals are reared. The more I found out about the treatment of factory farmed animals, the more difficult I found it to justify eating meat.

The last 40 years have seen the principles of mass production introduced into the farmyard. In the name of the great god efficiency, production systems have been devised that inflict unspeakable suffering on calves, pigs, chickens and turkeys throughout their short and miserable lives.

- Pigs are locked in crates, where they can only stand or lie and where they suffer great stress. They are bred to unnatural sizes and forced to produce more than 20 offspring a year.
- Broiler chickens, reared for their meat, never see daylight. They are crammed together in huge windowless sheds that are littered with excrement, which is never removed in their 42-day lives. They are forced by selective breeding to grow from chicks to adults at twice their normal rate of growth to the point where they can barely walk.
- Hens, reared for their eggs, are confined for their entire lives to battery cages so tiny they cannot stretch their wings, let alone walk.
- Cows are selectively bred to produce much higher yields and are milked to exhaustion, with their swollen udders forcing their legs apart. They are often in great pain from mastitis and separated from their calves within a few days of birth. And what a terrible fate awaits their calves – or at least it did until the collapse of the beef industry – in the veal crates of France and Holland.

- Sheep have had their breeding patterns manipulated in response to the demands of supermarkets, so that lambs are produced in winter and not in spring, with the result that hundreds of thousands die of exposure.

What are we to make of the appalling mutilations inflicted by factory farmers on their animals? Lambs castrated either by removing their testicles with a knife or by using a rubber ring to cut off the blood supply so that the testicles drop off. Chickens and turkeys de-beaked with red hot blades. Piglets have their tails removed with red hot irons. Some even have their teeth clipped. All these medieval tortures are allegedly made necessary by the need to stop these unfortunate animals turning on each other in the frustration caused by their close confinement.

How can this be morally justified? The truth is it cannot. Agribusiness can only get away with it by hoping that consumers do not find out or that, if we do, we avert our eyes.

Unfortunately, none of the main political parties – my own included – takes farm animal welfare sufficiently seriously. Presumably because farmers have votes and animals do not. However, a growing number of people, particularly the young – who are otherwise disillusioned with politicians – care passionately about animal welfare.

Even people who are oblivious to the suffering of animals have begun to notice the potential threat to their health and to that of their children. Bovine spongiform encephalopathy (BSE) was not an isolated disaster, to be resolved by a few practical measures and the shelling out of huge sums of public money to compensate an industry which has to a large extent brought the disaster on itself. BSE was nature's revenge on the factory farmers. Rather than continuing down the same old road, we should view BSE as an opportunity for everyone concerned – producers and consumers alike – to learn lessons of a more general nature than any that have so far been suggested.

A NEW APPROACH TO AGRICULTURE

We need a wholly new approach to agriculture in general, and meat and dairy production in particular. We need to restore morality to an industry which has lost its way. We must wean agribusiness off factory farming, away from the overuse of antibiotics and routine barbarity. We need a system of subsidy that encourages good practice and not merely discourages, but actively penalizes, bad practice.

We must harness the power of the market, through quality assurance and labelling which clearly describes the method of production, enabling consumers to choose how they want their meat reared. The free marketeers are always talking about choice. We need some real choice. We need to know how our meat was reared and what chemicals it was fed.

We live in an age where market forces are everywhere triumphant. Certainly we should continue to resist, by proper regulation, the worst excesses of the market, but where possible we should work with it. If the big supermarket chains, whose demands are to a large extent responsible for the degeneration of animal husbandry, were to decide tomorrow that they would no longer stock factory farmed meat, the whole grisly industry would be transformed overnight. The recent decision by Iceland supermarkets not to stock GE food is a very useful pointer to the way forward.

The picture in the UK, at least, is not wholly bleak. Within weeks of taking office, the new Labour government played a leading part in persuading our EU partners to sign up to a Protocol annexed to the Treaty of Rome, which for the first time recognized that animals are sentient beings, capable of feeling pain, and not merely agricultural products. We must now work to give practical expression to these sentiments.

The new Food Standards Agency will, in future, ensure there is a clear distinction between the interests of the public and those of the producers and distributors. This was not always apparent when all responsibility lay with the Ministry of Agriculture, Fisheries and Food (MAFF).

There is, however, one ominous development which threatens all the good work of recent years. GATT and the WTO are perhaps the biggest threats to farm animal welfare today. Under GATT, as it is presently likely to apply, all that counts is free trade. All other values – animal welfare, the environment, social justice – come a poor second. In the name of free choice, we will have no choice.

In Europe, at least, governments have, belatedly, become aware of the threat to civilized values posed by GATT. Work is now going on to address the problem, but it needs to be given much higher priority. GATT and the WTO must become a priority for the farm animal welfare movement.

The other area on which to concentrate is quality assurance. If we insist on certain minimum standards for meat and other farm produce imported into the EU, then those who want to export to us will have to clean up their act.

There is a wider issue that concerns the survival of the human race – or at least a fair proportion of it. That is whether the huge and growing demand for meat makes sense in purely practical environmental terms. In China alone, demand for pork and chicken – the very meats which most lend themselves to factory farming – has grown massively. In India, poultry consumption is increasing by 15 per cent a year. Bad news for animal welfare. Bad news, also, for the global food situation. Factory farmed pigs and chickens eat grain. Already more than one third of the world's grain is used to feed livestock. As a way of feeding billions it is highly inefficient. China, which used to be an exporter of grain, now imports it. It also uses up a lot of water. Lester Brown of the Worldwatch Institute has calculated that it takes around 180 litres of water to produce one battery egg and around 25,000 litres per kilogramme of beef. I find these figures astonishing and I suspect many people will. If they, or anything resembling them, are accurate, how much water does it cost to rear the 20 million more pigs born each year in China?

It is our job to make people think seriously about these great issues and then to persuade those who hold political power to take appropriate action.

7 An Agenda for Reform: Farm Animal Welfare in the European Union

Mark F Watts

The massive one million-plus signature petition by CIWF to the European Parliament in 1991 eventually led to the unanimous adoption in June 1997 by the 15 member states of the EU of a new Protocol in which animals are recognized at last as sentient beings. But this historic achievement is not enough. The campaign for animal welfare must continue and focus on turning those words into real action.

The next challenge is to reform the CAP once and for all by making animal welfare the cornerstone of a new approach to farming in Europe. Real progress has been made on this:

- According to Agricultural Commissioner Franz Fischler, 'For the European Union ... the improvement of animal welfare is a very important political objective.'[1]
- Today, in all member states, there are binding animal welfare rules with minimum requirements for the keeping of laying hens, calves and pigs.
- Veal crates are being phased out.
- There are new Directives limiting the length of time for animal transport and export refunds are now to be made conditional on animal welfare provisions being respected.
- Recently, a new proposal on battery hens has been made and we are currently discussing minimum standards for vehicles for transporting animals.
- In January 1999 the European Parliament voted to phase out battery cages.

These measures are welcome but none have gone far enough. However, in all of these measures animal welfare organizations such as CIWF and the European Parliament have led the way. We must now lead the way again and argue that a more fundamental approach is now not just desirable but absolutely necessary.

THE CAP

The CAP and its predecessors once fed a hungry Europe rebuilding itself out of the ashes of the war. Circumstances have changed, yet the CAP continues largely unreformed. It now constitutes one of the world's greatest manufactured disasters. It is bad for animals but also bad for farmers, consumers, taxpayers, the developing world and the environment too.

Bad for animals

The report jointly released by CIWF Trust and the World Society for the Protection of Animals (WSPA) vividly demonstrates how by the system of quotas, subsidies and cash payments the CAP has systematically encouraged factory farming which inflicts a massive amount of suffering on animals.[2]

Bad for farmers

Some 80 per cent of the CAP is spent subsidizing just 20 per cent of Europe's farmers. According to the Commission, Europe has witnessed a sharp decline in the number of farms and in the number of people employed in agriculture. Less than 6 per cent of the workforce now work in agriculture in the EU. The long-term trend is a further drop in the number of farmers at a rate of 2–3 per cent per year. Such simple statistics hide the human consequences and the impact on the viability and vitality of rural areas.

According to the Commission in March 1998:

> *The support it (the CAP) provides is distributed somewhat unequally and is concentrated on regions and producers who are not the most disadvantaged.*

This is having negative effects on regional development planning and the rural community, which has suffered badly from the decline in agricultural activity in many regions.

Bad for the environment

The European Commission has admitted that the encouragement given to irrigated crops through the regionalization of direct payments to cereals, oilseeds and protein crops, as well as relative advantage given to intensive livestock farming through lower feed prices and subsidizing silage, have all had negative effects on the environment.

On 18 March 1998, the Commission stated in the proposed CAP regulations that, 'other regions have seen the development of excessively intensive farming practices which are having often a serious impact in terms of the environment and animal disease.' Recent reports by the National Trust and WWF have documented examples of the CAP flying in the face of the EU's own environmental objectives. For example, in the UK, severe overgrazing and damage to very important habitat is occurring at the Long Mynd (Shropshire) as a result of headage payments. In Spain and Portugal, heavy incentives for cereal growing which requires intensive irrigation have caused chronic over-abstraction from aquifiers, threatening wetlands of Special Areas of Conservation status.

Bad for the consumer

The CAP adds about an estimated £20 a week to the average British family food bill. Beef and cereal prices, in particular, are way above world prices. Moreover, food safety has deteriorated and public confidence is at an all time low. This is not only because of BSE, with costs likely to exceed £4 billion, and new variant Creutzfeld-Jakob disease (nvCJD), which has led to the death of over 20 people, but also the increasing incidence of food poisoning which has increased fourfold in just over a decade. The intensification of agriculture has, at the very least, been a contributory factor in explaining this deterioration in food safety.

Bad for taxpayers

The CAP costs 40 billion European Currency Units (ECU) a year, which has to be financed, and future projections expect the figure to rise, not fall. No other industry receives such support from the taxpayer. Indeed the EU is now determined to tackle state aids in other sectors, but not apparently this one.

Bad for the developing world

Our surpluses are dumped on the world markets, depressing prices for other producer nations. They contribute to dependency problems and undermine attempts to develop sustainable development.

TIME FOR REFORM

The Commission has summed up this argument for reform:

> *All these factors combine to create a bad image of the CAP in the minds of the public. An agriculture which pollutes, which contributes inadequately to spatial development and protection of the environment, and which, because of its undesirable practices, must take its share of the responsibility in the spread of animal diseases, has no long term survival and cannot justify what it is costing.*

Despite the welcome rhetoric of reform, in reality the 40 billion ECU monster will continue to grow, threatening to feed on Europe and the rest of the world.

The time for radical reform is now, with the prospect of enlargement of the Community to Central and Eastern Europe, financial reform of the EU institutions and GATT forcing the pace. There are a number of reasons:

- The CAP takes around half the entire EU budget. The Agenda 2000 reform proposals indicate that by 2006, 50 billion ECU will be spent on market support with only 2.1 bn ECU on new rural measures in addition to 2.8 bn ECU on the three existing agri-environment, early retirement and afforestation measures. Even worse, these new compensatory direct payments are open ended and unconditional.

- Despite the desire of Commission President Santer to convince member states at Amsterdam to apply co-decision to Article 43 of the Treaty – all farm policy decisions – he failed. The Treaty Provisions relating to agriculture remain almost unchanged since the CAP was set up 37 years ago! This means the European Parliament only has limited power. CAP is only subject to the consultation procedure. We are allowed just one reading [of proposals] and even then the Council of Ministers can totally ignore what we say.
- CAP cannot be exported to other countries of Central and Eastern Europe currently applying for membership. For example, within the EU, agricultural employment represents around 3 per cent of the labour force. The figure for the accession countries is 24 per cent. We simply cannot afford to apply the CAP in an enlarged Union.
- The next round of multilateral trade talks under the WTO is set to commence at the end of 1999. Cutting border protection, reducing export subsidies and reshaping internal support towards more 'decoupled' instruments will all be necessary to meet the likely pressures required by the forthcoming GATT round. Lack of action will result in increasing levels of non-exportable surpluses after 2000, particularly for beef, cereals, sugar, wine, olive oil, skimmed milk powder and some other dairy products.

Our current policy will simply not survive the worldwide trend towards free trade. New Zealand now farms on a free-market basis and US agriculture also now works without production-linked support.

AGENDA 2000

As a result of these challenges, the European Commission published Agenda 2000 in July 1997. It attempts to set out a new approach and there is no doubting the commitment of EU Agricultural Commissioner Franz Fischler to the concept of decoupling payments from intensive production and linking payments to good animal welfare. Commission President Santer also stated in a speech to the European Parliament that any starting point for CAP reform must focus on animal welfare and a return to more natural production methods.

The National Farmers Union (NFU) in the UK would like CAP reform to go much faster and further. 'If our analysis is correct it will be extremely difficult for the EU to avoid further, possibly far reaching changes to the CAP early in the next decade.' They add that, 'It is at least arguable that it would be less disruptive to farmers and the industry as a whole to have a single major reform which will set the direction for many years to come.'

Lord Plumb, a Tory MEP and former NFU President, recently declared that, 'In reality my role is more that of a guardian of the land than a farmer.' Unfortunately, despite these high-minded sentiments, the reality is rather different. When Commissioner Fischler published his detailed proposals for CAP reform in spring 1998 they faced bitter opposition from Ministers within the agriculture council. In particular, for many of them, animal welfare concerns are clearly peripheral, and they are not facing up to other fundamental issues, such as world trade rules. In addition, price support and export subsidies will continue to consume over 90 per cent of the CAP whilst rural and structural policies will continue to make do with less than 10 per cent. The system of compensatory payments to cushion the decrease in price support will cost an additional 8 billion ECU by 2006 with very few additional resources to develop an integrated rural policy.

Payments are also planned to compensate farmers in perpetuity, and many of the welcome moves to encourage extensive production and compliance with environmental rules will be left to member states. In particular, each member state will be able to strike the balance it wants between intensive and extensive production.

This is hopelessly inadequate. Only a more radical approach will do.The replacement of the CAP with an environmentally sustainable food and rural policy is perhaps the only way of meeting public spending limits and international trade obligations, while rejuvenating rural life, enhancing the environment and safeguarding the welfare of animals.

An agenda for reform

The main components of this more radical agenda for reform are:

- 'The need to ensure high standards of health and welfare for farm animals' should be inserted in Article 39 as one of the objectives of the CAP.
- Article 43 should be amended to allow co-decision on all farm policy matters [ie this will allow majority voting].

- The CAP budget should be subject to complete democratic accountability and control by the European Parliament.
- Compensatory payments must be phased out by 2006.
- In the interim, all direct payments must be dependent on farmers meeting basic animal welfare and environmental standards.
- Decouple payments and phase out quota payments which encourage intensive livestock production.
- Support for low intensity and traditional arable and livestock farming.
- Support for organic farming.
- Switch resources to develop agri-environmental measures and rural development programmes.
- An immediate end to export refunds which encourage live animal exports to third countries.

Animal welfare must not simply stop at the frontier of the EU. We must continue to work for better rules at international level as well. Only a move away from factory farming to more extensive systems of production is capable of squaring the CAP reform circle. A reformed CAP will only work if it is enforced uniformly throughout the Community and the Community acts as a power for good, enforcing high standards throughout the world.

I think the citizens of Europe will not tolerate a situation where we agree legally binding rules and they are ignored by the member states. Some member states are clearly better than others. We have got to give the EU more resources to properly police its own rules, because even if we reform the CAP, if high animal welfare standards are not enforced it will be meaningless.

More generally, if we are to achieve such a reform, animal welfare campaigners have got to find support from a broader coalition of people like taxpayers, consumers, those concerned about the environment and farmers. Not all farmers, clearly, but some of them who are being driven out of business by the CAP. Most taxpayers do not know what happens to their EU contributions. When I tell them what is spent on the CAP most are shocked and disgusted. They do not believe they should be subsidizing agriculture in this way and then of course paying twice through the shops because of price support. We have got to win over allies like taxpayers and consumers to say quite simply that they are not prepared to carry on paying for this policy.

Notes

1 28 November 1997, Brussels
2 Michael Winter, Charlotte Fry and Peter Carruthers (1997) 'Farm Animal Welfare and the Common Agricultural Policy in Europe – A Report to Compassion in World Farming Trust & World Society for the Protection of Animals', CIWF Trust/WSPA

8 Campaigning for Change in the European Union

Philip Lymbery

In recent decades, the Western world has seen the rise of intensive farming. During the same period, we have witnessed perhaps the greatest disappearing act of our time:

- Our wildlife has disappeared and the richness of our natural environment has drastically diminished. Populations of once familiar farmland birds have seen drastic declines. Many miles of hedgerows have been ripped up and our wildflower meadows and other natural habitats have been disappearing under the plough.
- Farmers, too, have been disappearing and rural communities taken to the brink of collapse. Between 1946 and 1989, the total number of people working on farms in the UK (full and part-time) declined from 976,000 to 285,000.
- But perhaps the most marked disappearance has been seen in our farm animals. They have practically disappeared from the land, only to be caged, crammed and confined behind the closed doors of the factory farm. Here they are treated like animal machines, like Cartesian automata – devoid of all feeling. They are reduced to the level of mere commodities, goods, agricultural products.

In Britain alone, 30 million hens are imprisoned for life in battery cages so small that they cannot even stretch their wings; 700 million broiler chickens are reared each year by being crammed in their thousands into windowless sheds. They are made to grow so fast that their bone structures cannot keep pace, often resulting in painful crippling. Eighty per cent of our breeding pigs and an even greater proportion of those fattened for slaughter are kept in barren and cramped factory farm conditions. Ninety per cent of the EU's dairy cow herd is made up of Holstein Friesian cows, the product of the overzealous breeder. These animals are often kept in inadequate housing and are pushed to their physiological limits in the quest for ever higher milk yields. Today, the vast majority of our farm animals in the UK and Europe are factory farmed.

Animals seem to be regarded as 'cogs in a machine' in the quest for what some narrow mindedly call 'efficiency'. But what is efficient about a system which causes great environmental damage, spectacular public health disasters such as BSE, or immense suffering on the part of farm animals? It is a definition of efficiency based on a barefaced acceptance that someone, or something, else will pay the real cost.

THE CONCEPT OF WELFARE POTENTIAL

In the UK, intensive agriculture is increasingly coming under fire from environmentalists, animal welfarists and an ever more questioning general public. The industry fights back with superficial arguments, tired, familiar lines about the UK's farmers 'having the highest welfare standards in the world'. Yet millions of animals still languish in close confinement, or suffer pain from being souped-up by selective breeding. We are asked to believe that high standards of stockmanship are all that is needed for good welfare. Of course a high level of stockmanship is important, but its effectiveness is wholly limited by the welfare *potential* of the husbandry system used.

The classic example is the battery cage for egg-laying hens. Even with the very best stockmanship in the world, the welfare of hens in a battery cage will still be poor. The cage's cramped and barren environment restricts and denies the birds' behavioural and physiological needs, and they suffer as a result. In short, the battery cage is a system with low welfare potential. No matter how much stockmanship and care you lavish on the birds in that system, their welfare will still be poor.

A free-range system, however – with its space and enriched environment – has a high welfare potential. Of course, if the level of stockmanship is poor or neglectful, then the birds will suffer. But then, high standards of stockmanship should be a must, not an option, in any farming system. Similarly, a badly designed unit would also have a negative effect on the birds' welfare. The point, though, is that any problems of design or husbandry technique in these free-range-type systems can be ironed out and the full welfare potential of the system achieved.

CAMPAIGNING

CIWF has been campaigning to end factory farming in the UK and Europe for 30 years now. Our campaigns work on three different levels:

- the public campaign, aimed at raising general awareness, stimulating public concern and encouraging people to use their power as consumers to help bring about change;
- the political campaign, to follow through and consolidate our public work by achieving concrete legislation to protect the welfare of farm animals; and
- the campaign to influence policy mechanisms, such as the CAP, to help encourage better welfare on a Europe-wide basis.

Public campaigns

CIWF was set up by Peter Roberts – a man of great courage and vision. In the 1960s, Peter was a dairy farmer, who also kept poultry. He started his campaign out of concern at the growing use of systems such as the battery cage for hens and veal crates for calves. Thirty years on, the battery cage has been condemned on welfare grounds by a succession of official bodies. These include the UK House of Commons Agriculture Select Committee (1981), the European Parliament (1987 and 1999) and the European Commission's own Scientific Veterinary Committee (1992 and 1996), which concluded that battery cages have 'inherent severe disadvantages for the welfare of hens'.

Battery cages

Although 89 per cent of the British public now believe it to be cruel to keep hens in these cages, 85 per cent of Britain's hens are kept in this system in the UK. So is the general public saying one thing and doing another? Not necessarily. Until recently you could not buy battery eggs anywhere! What you could buy was 'farm fresh' eggs or 'country fresh' eggs or simply 'fresh' eggs – all labels used to sell battery eggs. These labels have been shown to mislead a significant proportion of people to the point where they think those eggs were actually laid by free-range hens. Recent opinion polls show us that more than a third of people are fooled by these labelling terms.

Thanks to hard campaigning, some supermarkets are adopting clearer labelling policies, using the terms 'eggs from caged hens' on their battery eggs. Unfortunately, the vague term 'fresh' eggs still predominates. Unclear labelling hampers true consumer choice.

Even though 85 per cent of the UK's egg-laying flock are in battery cages, it does not follow – as has been commonly believed – that only 15 per cent of shoppers are buying non-caged hens' eggs. A CIWF survey of leading supermarkets on egg sales in spring 1998 found that in three of the top five major multiples between 37 and 52 per cent of their total sales of shell eggs were from alternative systems such as barn or free-range. Most other supermarket respondents declared that alternative eggs make up 25 per cent or more of their total egg sales. In one notable case, Marks & Spencers, 100 per cent of eggs bought are now free-range. This shows a much greater level of consumer action on this issue than was previously thought.

A large number of battery eggs are used either in liquid form for processing, or in catering, where the system of production is not declared on the final product. Add those to the proportion of potentially confused consumers over labelling, who usually think they are buying non-cage shell eggs, and we can see that the consumer movement towards kinder ways of keeping hens is considerable.

CIWF is working with kindred societies throughout the EU to ensure that the public and political condemnation of battery cages is consolidated in an urgent phase-out of this system now that the review process of the Battery Hens Directive has finally begun.

Veal crates

A striking example of a successful public campaign is that against veal crates for calf rearing. As a dairy farmer in 1967, CIWF's founder, Peter Roberts, began speaking out against the veal crate system. It was a campaign which sparked off probably the most successful consumer boycott this country has ever known. Watching their TV sets, reading their newspapers, listening to their radios, people were horrified to find that tiny calves were kept in narrow solid-sided wooden boxes, so narrow that the calves could not even turn around. They were shocked at the all-liquid, iron-deficient diet used to produce the white veal from anaemic animals so prized by gourmets.

The campaign was so successful that when people were out shopping, they would pick up the white veal, remember what they had seen or read, and put it back on the shelf. They did not want to buy this cruelly produced product. The boycott proved so successful that by the time the government acted in 1987 to phase out this system, there were only eight veal crate farms left in the UK. The others had either switched to a more humane system, or had gone out of business. The veal crate was finally banned in the UK in 1990. It is a tremendous example of how consumer power can bring about real change. It also demonstrates how public opinion and political progress are so often inextricably linked.

Consumerism and citizenship

We are all, to some extent, consumers. We do have the power of our purse to help make our views known on a range of issues. But at the same time, we are all citizens. Citizenship brings with it rights and duties. One of those rights is to proper representation within our democratic process.

People as citizens are instinctively offended in moral terms by the way animals are treated on factory farms, as the response to the veal crate shows. Then, consumerism and citizenship acted together towards the same goal – an end to the cruelty of the veal crate.

However, for many of us, consumerism and citizenship are often not operating at the same time. This is partly because when we are out shopping, our minds are focused on just that – shopping. Moreover, ethical consumerism is not easy! In today's modern stores we can be surrounded by an array of competing products labelled, of course, in a way which promotes what is good – the selling points – about the products. When talking to supermarket insiders, it is clear that some shoppers tend to feel that supermarkets would not stock or sell anything which was in any way ethically substandard. In addition, shopping often takes place in a carefully controlled environment full of reassuring messages. Ethical consumerism is not easy when a simple statement such as 'fresh' on farm produce is often interpreted as a statement on the humaneness of production rather than simply denoting newness.

People may not be experts. They may not know the difference between 'fresh' eggs and 'free-range'. But they are citizens. They do have a sense of what is right or wrong. It is the duty of elected government to realize our aspirations as citizens. Time after time, people as citizens show deep-seated concern at the way animals are treated. It is up to politicians to ensure those concerns are recognized and acted upon. Legislation is a legitimate method of meeting those concerns and protecting the welfare of farm animals, which is so often woefully inadequate.

I am sure that, as citizens, most people in the UK now look to our elected representatives to ensure they do not fail on the battery cage issue, but instead consign this system to the scrap heap of history.

Live exports

Citizens also protest about what they do not like. In 1995, the power of popular protest was seen at infectious demonstrations taking place at ports and airports around the country. Here, there was no real consumer angle. Ordinary people – men, women and children – of all ages, all classes, all professions, came together to voice their anger at the export of live calves and sheep from Britain to the continent. At that time, about two million sheep and lambs per year were being taken off the hills of Scotland and Wales, driven from the valleys of England, forced to endure the hustle-bustle and confusion of the livestock market before being packed into poorly ventilated lorries. On the dockside they would await an often rough sea-crossing before the horrifically long journey to a far-away abattoir. Journeys could often last for 30 hours or more, without food, water, or rest. Half a million calves, too, were being exported to end up in narrow veal crates on the continent.

Sadly, the live animal export trade still continues today. So did these protests achieve anything? The answer is a simple and emphatic *Yes*.

The protests brought about a dramatic drop in the number of sheep being exported. But much more than this, the protests put farm animal welfare firmly on the public and political agenda. Two tangible results were the monumental decision by the EU to phase out veal crates altogether, and the adoption of a Protocol giving legal recognition to the fact that animals are sentient beings – capable of feeling pain and suffering.

We have been tremendously successful in reawakening the concept that animals have their own intrinsic value. Farm animals may be kept for the food they produce for humans but they should not simply be viewed as a means to an end. As Schumacher pointed out, 'The production of food for humans is the animals' secondary, not their primary, nature. Before everything else, they are ends in themselves.' Farm animals are living, sensitive creatures with an incalculable value in their own right, and as such demand our compassion and respect.

Consolidation by legislation

Popular concern for the welfare of farm animals has already resulted in some legislative change. But there is still much more to do. Legislation is urgently needed to end the factory farming of pigs and poultry for meat. Different systems have different welfare potentials. Apart from the housing system, discussed above, two other major factors limit this welfare potential – the breeding and the feeding regimes employed:

- In the production of broiler chickens for meat, all three factors play a part in an extremely low welfare-potential system. Thousands of broiler chickens are crammed into barren, windowless sheds. They are forced to grow at phenomenal rates due to selective breeding and the feeding of high-protein diets which usually include growth-promoting antibiotics. Each bird is only given about 0.5 sq ft of floor space, an area smaller than the cover of a telephone directory. They grow so fast that their skeletal systems cannot keep up with their superfast-growing bodies. One quarter of broilers are estimated to suffer from leg problems that leave them in chronic pain for 25–30 per cent of their lives. They are usually slaughtered at just six weeks old. The young slaughter age does not prevent 17,000 broiler chickens in the UK dying from Ascites-related heart failure every single day. 700 million broiler chickens are reared and slaughtered every year in the UK alone. The number of these birds being produced is increasing, not only in the UK but also throughout the EU and on a global scale.
- Pigs, too, are in desperate need of legislative protection. Most of the UK's 13 million fattening pigs are kept in barren, cramped conditions and forced to endure floors of concrete or slats devoid of bedding. Boredom and frustration result in these animals biting each other's tails, so the farming industry resorts to the violence of tail-docking – cutting off the top of the tail, using a hot pair of clippers or pliers. Despite the routine docking of piglets' tails being banned in the EU since 1994, 70 per cent of our piglets still endure this painful mutilation. Pigs have also been selectively bred for faster growth rates and greater muscle development. They, too, are increasingly suffering from painful joint and leg problems and are also more susceptible to heart strain.

CAMPAIGNING TO CHANGE THE CAP

As well as campaigning for legislation to ensure that if you are cruel and break welfare laws you will be properly punished through prosecution, CIWF also campaigns for policies to encourage good practice. A key area is influencing policy mechanisms such as the CAP to encourage kinder, more humane farming practices.

The CAP has encouraged the industrialization of agriculture. It has fanned the flames of farm specialization, of single-minded productivity at almost any cost. It has encouraged a consequent decline in traditional mixed farming and is partly to blame for the great disappearing act of animals from the land.

The CAP, without doubt, has helped to make Europe's cattle into welfare victims:

- Modern dairy cows in the EU suffer the double insult of inadequate cubicle housing and over-demanding production. Milk quotas aimed at curbing oversupply from an industry in overdrive have encouraged a push towards ever higher milk yields from individual cows. This has imposed metabolic and physiological stresses on the animals. The modern dairy cow simply does not have the metabolic capacity to keep up with her overproducing udder, resulting in a host of welfare problems such as mastitis, lameness and early culling.
- In the UK alone, over one million male calves have been slaughtered at less than 21 days old, under the so-called 'Herod' scheme. Generous CAP subsidies of about £100 are paid for each calf slaughtered under the Calf Processing Aid scheme. The meat is not allowed into the human food chain, not because of any BSE-related fears, but to reduce the EU's beef surplus.
- CAP subsidies are also responsible for encouraging perhaps the cruellest aspect of the live animal export trade – cattle exports to the Middle East. Export refund payments of around £400 per animal are made to exporters of live cattle to non-EU countries. Some half a million live cattle were exported from the EU in 1996 alone. The long journeys involved in this trade inflict immense suffering on the animals. In addition, they are delivered to countries where animal welfare standards are almost non-existent.

Around £200 million of CAP money per year is spent on directly subsidizing this trade. But CAP funds are public money, yours and mine, being used to pay for tremendous cruelty. Almost half of the EU's total budget of nearly £60 billion (1996) is swallowed by the CAP.

It is surely time for reform. Reform of a system which makes such victims of cattle, and where pigs and poultry are considered a mere offshoot of the cereals sector. CAP spends taxpayers' money but takes no account of welfare or the intrinsic value of animals. It is time to reform it:

- to ensure that public funds should only be payable where high welfare standards are met;
- to encourage a return to more welfare-friendly mixed farming methods;
- to help the development of organic farming within the EU; and
- to ensure that schemes which encourage cruel practice are abandoned.

In reality, the whole outlook of the European farming industry needs reform. As part of a global movement for compassion and sustainability, we in the EU can play a major part by turning the tide away from the failed policy of industrialized agriculture. Together, we can also help stop the flow of often inappropriate and damaging factory-farm technology to developing countries. The new direction for European agriculture should now recognize the sentiency and intrinsic value of farm animals, protect the environment and maintain rural communities, both for today and tomorrow.

9 Farm Animal Welfare in Central and Eastern Europe

Janice H Cox

Central and Eastern Europe is a large and diverse region. Some countries in the region are reminiscent of Western Europe, some are more Southern European in character, and some are more akin to developing countries. The farming in these countries varies greatly – from predominantly industrialized agriculture to almost entirely small scale, private farming. Such diversity makes it difficult to cover this subject coherently in a short chapter. However, there are immense farm animal welfare problems in the region caused largely by communist ideology. Some key changes have affected farm animal welfare since the fall of communism – especially in the five Central and Eastern European Countries (CEECs) which are in the first wave for EU membership (the 'first wave countries'): Poland, Hungary, the Czech Republic, Slovenia and Estonia.

The Communist era

Communist ideology destroyed man's relationship with animals. When the Communists gained power, they repealed pre-war animal protection laws and banned animal protection societies in most countries. Farm animals were taken out of their pastures and barns into large, cramped industrial units. Over the socialist period, the combination of industrialized state and collective farm structures and negative attitudes bred immense welfare problems, which were exacerbated as economies and infrastructure deteriorated.

Farm animal welfare problems

Five problems made a major contribution to poor farm animal welfare during the socialist period:

1. *Lack of legislation/control.* There were no comprehensive animal protection laws. When comprehensive animal protection laws were repealed, this left just some old Soviet laws containing broad anti-cruelty provisions, but these were rarely enforced and penalties were low. Official checks, such as veterinary supervision, covered health and hygiene, but not animal welfare.
2. *Industrialized farming.* The vast majority of farm animals in CEECs were raised in state or collective farms, usually in indoor systems with zero grazing. The sheer scale of these units, coupled with increased mechanization, broke the relationship between stockman and animal. There was no sense of ownership and responsibility. Production was centrally planned, with inputs allocated and output predetermined.
3. *Poor stockmanship.* Work with animals was considered low-level employment. There was a lack of training for stockmen, and no animal welfare training or awareness. The system destroyed the work ethic, professionalism and job interest. Work avoidance became the order of the day. Jobs were secure, and there were no incentives or disincentives.
4. *Slaughter/transport.* Although there were some export slaughterhouses in the region (especially to EU/USA), there were many more with antiquated facilities and unacceptable practices. Ineffective stunning – or no stunning at all – was common. The Soviet system favoured electrical stunning, even for adult cattle, but often the equipment used was ageing and poorly maintained. In rural areas, lack of stunning was common, and old methods – such as sledgehammer – continued to be used.

   Casualty slaughter caused serious welfare problems. Casualty animals were often left to suffer until routine work was finished. Even then, little effort was made to minimize suffering caused through movement and choice of killing method. However, one plus for the cooperative system was that animals were often slaughtered near to the point of rearing.
5. *Lack of awareness of animal welfare.* Animal protection societies were normally banned in the region. Some societies remained in Poland and former Yugoslavia, but in most animal protection was seen as an undesirable and insurgent activity. Students in further

education (including veterinary and agricultural institutes) and at school were discouraged from discussing animal welfare. Moreover, the media were not free, and animal welfare issues were rarely covered.

POST 1989

Agricultural reform

With the fall of communism from 1989, the process of agricultural reform began in order to 'return to private farming and private ownership the factors of production'. The main objectives of agricultural reform were to:

- de-collectivize and privatize agriculture; and
- re-establish property rights.

Different routes were taken in different countries, for example:

- A mixture of restitution and compensation in some, such as Hungary, the Czech Republic and Slovakia. This involved state farms being privatized, and old cooperatives being converted to private businesses. Where land could not be returned to former owners, they received compensation, sometimes in the form of shares.
- Liquidation and restitution, as in Bulgaria. In this case, all state farms were liquidated and there was almost full restitution to former owners. In Romania and Albania, former collective farms were closed and dismantled.

Thus, whilst previously the vast majority of farms in the region were collective or state, this situation has changed dramatically with about four out of every five animals now owned by private farms. The situation in these 'first wave countries' is shown in Figure 9.1

Trade changes

With the fall of communism in the region, old socialist trade ties (Comecon) were broken, for example:

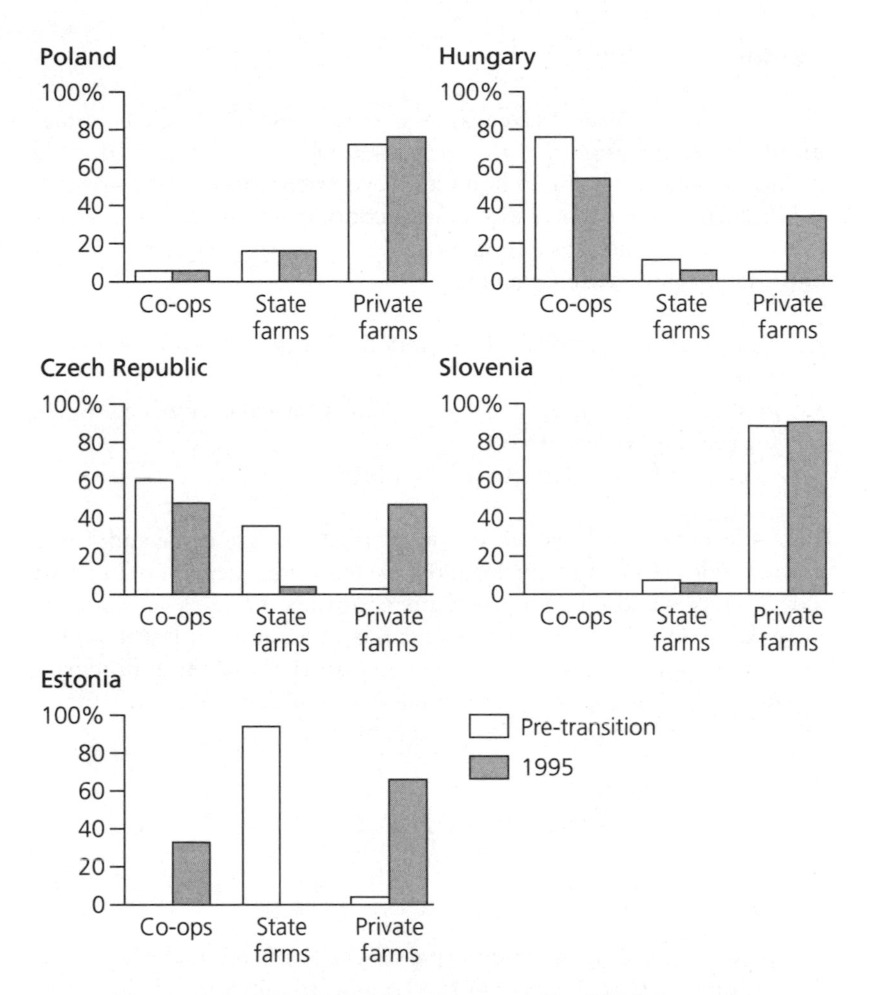

Figure 9.1 *Change in Farm Structure in CEEC Countries*

- The Baltic countries' market for livestock was previously Russia.
- Many CEECs benefited from cheap Soviet energy and animal feed.

They then had to develop new export markets and to buy their imports at world prices. For many CEECs, the EU is now a major trade partner, and live animals and meat are among the region's major exports to the EU.

Sector contraction

From 1989 to 1994, there was a considerable decline in animal numbers in the region. Initially, this decline was largely due to producer' sales to maintain liquidity. These were followed by substantial liquidations due to the worsening economic situation, and changes in production structures and markets. The 'price-cost squeeze' was an important factor, caused mainly by:

- almost universal removal of food subsidies (raising food prices and lowering demand);
- general deterioration of economies (weakening consumer purchasing power); and
- steep input price rises (to world levels).

Most affected by the livestock sector contraction were cattle and sheep in the original CEFTA countries (Poland, Hungary, Czech Republic and Slovakia), pigs and sheep in the Baltic countries and cattle and poultry in Balkan countries. The dairy sector has been most resistant, whilst pig production has remained high in countries with high domestic consumption, such as the Czech Republic and Poland. The situation in the 'first wave countries' is shown in Table 9.1.

EUROPEAN INFLUENCES

EU

The prospect of EU membership (possibly towards the end of the next decade for first wave countries) is a beneficial influence. In principle, countries have to implement the entire body of EU law (the 'acquis') before entry, although in practice certain transitional periods can be agreed. The five CEEC applicants have already embarked upon the process of harmonizing their legislation with EU provisions. This process begins with the mammoth exercise of comparing national law with EU provisions, usually in some sort of priority order. Unfortunately, animal welfare is seldom viewed as a priority, unless it gives rise to trading implications. However, the process of comparing and harmonizing animal welfare laws has already started in all of these first wave countries. For example, the Czech Republic has set up a special group to consider animal welfare law harmonization, and has already introduced EU slaughter regulations (as well as a comprehensive animal protection

Table 9.1 *Livestock production decline in CEEC countries*

Country	*1989*	*1994*	*%*
Poultry production (thousands)			
Poland	66,188	53,330	81
Hungary	61,604	33,612	55
Czech Republic	32,479	24,974	77
Slovenia	13,279	10,592	80
Estonia	6932	3226	47
Sheep production (thousands)			
Poland	4409	891	20
Hungary	2215	1252	57
Czech Republic	399	196	49
Slovenia	24	20	83
Estonia	100	50	50
Cow production (thousands)			
Poland	4885	3866	79
Hungary	568	420	60
Czech Republic	1247	830	67
Slovenia	243	210	86
Estonia	300	227	76
Pig production (thousands)			
Poland	18,835	17,422	92
Hungary	8372	5001	60
Czech Republic	4685	4071	87
Slovenia	576	591	1
Estonia	1099	424	39
Cattle production (thousands)			
Poland	10,391	17,270	70
Hungary	1690	999	59
Czech Republic	3481	2161	62
Slovenia	546	478	88
Estonia	819	463	57

law). Estonia planned to compare its legislation to EU provisions (and to introduce a comprehensive animal protection law), and Slovenia has set itself a deadline of the year 2000 for harmonization.

In preparation for eventual EU membership, a series of association agreements, known as the 'Europe Agreements', have been concluded with certain CEECs. These and separate agreements covering 'sensitive sectors', including agriculture, incorporate trade liberalization measures. But relevant EU welfare criteria must be met for live animals and animal products to be exported to the EU.

Council of Europe

Most CEECs are now members of the Council of Europe (CoE), including all the first wave EU applicants. The CoE is the bastion of human rights in Europe, and aims to improve European unity, democracy and human values. The culture of the CoE is clearly beneficial to broader humane values. Amongst its conventions are five covering areas of animal welfare. Three of these are relevant to farm animals: the conventions on farm animals, slaughter and transport. Whilst ratification of these conventions is voluntary, some CEECs are clearly working towards this end. For example, Slovenia has already ratified the conventions on farm animals and slaughter, the Czech Republic ratified those on farm animals and transport in 1998, and plans to ratify that on slaughter. Estonia is comparing its legislation to the welfare conventions as well as EU law, with the aim of alignment.

National

In some cases, CEECs are willing to go beyond harmonization with EU and CoE animal welfare provisions to include additional national welfare provisions. Some countries in the region are introducing comprehensive animal protection laws that would put many Western European countries to shame! The Czech Republic was the first in the region, introducing a comprehensive animal protection law as early as April 1992, and by 1996 had trained 296 animal welfare inspectors (from within its Veterinary Service). Slovakia, Poland, Hungary and Estonia have also introduced animal protection laws, and other CEECs are poised to follow.

Joint Ventures and Foreign 'Expertise'

Another influence is increased contact with other European countries, both at business and personal level. Joint ventures with Western European businesses are common, bringing both capital and expertise into CEEC farm industries.

CHANGES BENEFICIAL TO ANIMAL WELFARE

European influences

The CoE and EU influences are of primary importance. The CEECs strive to demonstrate their open and democratic credentials in these fora, and some are already working to harmonize their animal welfare provisions.

Trading changes

Increased trade with the EU brings with it compliance with relevant EU animal welfare legislation (such as slaughter and transport).

Farm structure changes

Structural changes bringing smaller livestock sectors and increased private ownership are likely to be beneficial to farm animal welfare in the longer term. Owners will have a financial interest in caring for their animals, and employees will no longer have unconditional job security. Newly established private farms are already introducing outdoor access/grazing.

Awareness of animal welfare

Awareness of animal welfare is increasing. Veterinary and agricultural institutes are beginning to introduce animal welfare and behaviour into their courses. There has also been some training introduced for stockmen and slaughterhouse technicians, but this remains limited. Western European contact often imparts a positive welfare message. Some Western European joint venture partners bring their own standards of animal welfare to the venture, particularly Scandinavian companies. The animal protection movement in the region is growing in strength, and has ever closer contacts with well-established Western European and international societies. But unfortunately their work often centres on companion animals, despite the massive need for action in the field of farm animal welfare.

The free media also cover animal welfare issues increasingly (but less frequently for farm animal welfare issues).

CHANGES DETRIMENTAL TO ANIMAL WELFARE

Structural changes

The downsizing of the livestock sector in CEECs brought with it immense farm animal welfare problems as farms collapsed and could no longer afford feed and energy. Although frequently viewed from outside as a panacea for farm animal welfare problems, privatization has brought its own set of welfare problems. When farms are restored to private owners, these frequently do not have the financial resources, equipment, expertise or experience needed for livestock farming. Some come from an urban lifestyle, and view livestock farming as simply a new opportunity to make a profit. Clearly, it will take time before the private farming sector brings the expected welfare benefits.

Foreign contacts

Foreign contacts and business relationships are not always beneficial to farm animal welfare. Some foreign companies are seeking relocation and/or investment in CEECs in order to take advantage of the competitive edge brought by lower costs and lower animal welfare standards. Such relationships can be exploitative, rather than 'educational'.

Economic pressure/competition

There is a danger that new economic pressures in the region, including increased competition, will lead to 'corner cutting'. This could also have a detrimental effect on farm animal welfare.

FUTURE PROSPECTS

In my view, the future prospects for farm animal welfare in the region are that:

- The prospect of EU membership and CoE influence will raise farm animal welfare standards in CEECs. These fora should make special efforts to assist CEECs in this area.

- Structural changes, including private ownership, are likely to enhance farm animal welfare in the long term, but there will be disasters in the process.
- Increased trade and slaughterhouse closures will lead to more transport of farm animals. Training and higher vehicle standards are needed if welfare disasters are to be avoided.
- Veterinary and agricultural institutes will be of prime importance to the development of farm animal welfare in the region. Western European institutes with strong welfare programmes should make real efforts to assist their CEEC counterparts across the region.
- There is a real need to encourage animal protection societies in the region to campaign and educate on farm animal welfare issues. Western European and international societies could make a vital contribution to this process.

Bibliography

Agricultural Situation and Prospects in the Central and Eastern European Countries (1995) European Commission

Economic Situation and Economic Reform in Central and Eastern Europe. Economic Reform Monitor

Central and Eastern European Agriculture and the EU, Agra Europe (London) 1997

The Status of Animal Protection in Central and Eastern Europe. The World Society for the Protection of Animals (WSPA) (1995)

Animal Protection Situation in 1996, The State Veterinary Service of the Czech Republic

Personal correspondence and questionnaires from: the Central Commission for Animal Welfare, Czech Republic; Estonian Veterinary and Food Department (Ministry of Agriculture); the Veterinary Administration of the Slovenian Ministry of Agriculture, Forestry and Food; Prof Dr Laszlo Visnyei, Hungary; and Ewa Gebert, Polish SPA

Chart of ratifications, and personal correspondence, Council of Europe

Assorted EU publications and Internet information including *Background Report on the Europe Agreements* (1995); Agenda 2000; White Paper on Accession

Eurostat

OECD& FAO, statistics and various reports/information

10 Factory Farming and the Meat Industry in India

Maneka Gandhi

In Asia, home to over 60 per cent of the world's population, India is perhaps the only nation that concerns itself with animal welfare at all. We have laws on animal welfare and even attempt to follow them. Animals are also mentioned in our constitution as something more than just for eating. Each citizen is, in fact, enjoined to care for and preserve nature and its creatures. India is unique in Asia, but for how long? I am not overly optimistic about the future because there is this great belief sweeping in that everything the West does is right and we are beginning to imitate its worst mistakes. Junk food is fashionable. Eating meat is regarded as progressive. Modernization is equated with changing from being vegetarian to non-vegetarian even while the rest of the world attempts to reverse this trend. But in India eating meat is seen as a brave departure from tradition and old-fashioned values. Food processing no longer connotes dealing with vegetables or cereals or turning the earth to grow from it. Food processing is about meat and the meat industry.

As a largely agricultural nation, India is hugely dependent on its animals. Our farming systems rely almost entirely on them. However this does not translate into better treatment for these creatures, who are as badly treated as anywhere else. Apart from farm animals, we have meat animals – those raised for slaughter. Buffaloes and goats comprise the bulk of these. Sheep are raised for wool and killed for meat later. Then there is poultry: hundreds of millions of birds are killed every year.

Surprisingly, India only eats 25 per cent of these animals. The rest is exported primarily to the Middle East. This extraordinary statistic reflects the lopsided logic of killing our animals to feed other nations. Neither does it make sense economically. People wrongly believe that exporting meat earns the country money, when actually just the opposite is true.

To begin with, Indian meat is priced far lower than that from any other country – at only 40 per cent of world market prices – so we have to kill two to three times the number of animals to earn the same from meat export as any other nation. In addition, while the government does not give any money for loans for farm seeds and the like, there are a variety of subsidies and incentives offered by the Agriculture and Processed Food Export Development Authority (APEDA) for slaughterhouses and to promote exports of meat. The government has also agreed to subsidize with a 100 per cent grant nine slaughterhouses near airports. What we lose in terms of the economic contribution of these animals alive far outweighs what we gain from their death. Every dead goat, for example, fetches approximately Rs 250 (£4.00). But what we lose from its export amounts to Rs 60,000 directly and about Rs 100,000 indirectly. This is how it works.

Goat Rearing

Our goats are not grown on farms, in fact apart from poultry no animals are. People with land grow crops instead. People with no land raise animals. They are given pairs of goats as part of government relief programmes and self-help schemes. With no land to grow fodder for their animals, these goatherds raise and graze their animals on open ground – hillsides, parks and forest areas. Protected wildlife sanctuaries are the last green areas in India and amount to a paltry 8 per cent green cover for the whole country. But even this figure is dwindling because goat owners need green for their goats and this is where they find it. Each goat, according to a government survey, consumes ten acres of land before it is killed at the age of two. Nor does it simply eat the grass like a cow – it actually pulls up the plant by the root, destroys young shoots and leaves the earth bare. The topsoil flies off and pretty soon the land is barren. In Rajasthan, which is a desert to begin with, the goats destroy whatever little vegetation there is and encroach into sanctuary area all the time. The strong winds then scatter the sand from the ravaged area and the desert creeps up further. When the land becomes barren, the rivulets dry up. These little rivulets coming from the hills run into rivers that feed all of us. When they dry up, water levels go down and water shortages go up.

India suffers from an acute and perennial water problem, with states squabbling over river waters constantly. States that share river waters start political feuds and physical riots over getting less water than they need, but that is because there is less to go round. And that, in turn, is because there is less water coming into the rivers because goats are eroding the land and drying up its streams. So the raising and killing of goats leads to human conflict and killings as well.

Another irony is that though we have a great international campaign, launched with much fanfare and publicity, to save the tiger, this magnificent predator is being murdered due to goats. This is because when shepherds take their flocks into the jungle, they ensure the area is safe by killing all the tigers first. Meat poisoned with pesticide is placed near the watering holes and the unsuspecting tiger meets a most terrible end. In recent times, more tigers have died from poisoning than any other cause. In just one month, March 1998, we lost three tigers in the Jim Corbett Sanctuary, which is considered among the better protected reserves in the country. The shepherd is not interested in conserving the tiger, he is interested in conserving his goats. As long as goats continue to be raised for meat, the tiger is doomed to destruction.

Apart from killing the tiger, goats are also destroying all wildlife habitat. In the jungle, goats eat the undergrowth and all the young shoots, so no new forest comes up. Our forests are constantly under attack for fuel. This was always the case, only now the trees and branches that are cut are no longer replaced by new ones because of the herds that infest the area. About 15 years ago, our forests covered some 23 per cent of the land area. Today, we are down to a minimal 8 per cent, because you cannot have trees when you have goats.

Even city tree planting efforts fail for lack of tree guards. Why do we need tree guards? Because goats come and eat the plants. So, if the municipal authorities do decide to plant, they choose poisonous Oleanders or Astonia Calaris, which is known as the Devil Tree because no bird or animal will touch it. We have invented species called False Ashoka only because it will not be eaten by animals. So we have no trees coming up that produce either shade or fruit, simply because they need to survive the goats.

What does this all translate to in real terms?

The Moguls chose Delhi for the country's capital because of its perfect location. Surrounded by the Aravalli mountain chain and with its own river, the Yamuna, flowing right through it, it was a self-sufficient and secure city. Wild boar and deer were plentiful. People picnicked and hunted all the time. Today, Delhi is one of the most polluted and unhealthy cities in the world. The once dense Aravallis are totally denuded and the wind carries sand and dust into the city. Delhi has a very high SPM (suspended particulate matter) level and a correspondingly high level of respiratory diseases.

Over the last ten years, India has taken loans totalling $54 million from Japan to re-green the Aravallis. Even so, we have not been able to grow even a single blade of grass. Because the moment the grass appears, so do the goats. The Avarillis used to be a major catchment area for rain water. The vegetation used to trap the water, which used to seep into the soil and raise the water table. Now, in the absence of any vegetation, the water runs off rapidly without percolating into the soil. As a result, Delhi's water table has dropped drastically. This in turn has affected lifestyles. Rows of women must line up for hours at a stretch to fill a bucket of water. They must then choose between drinking and bathing and washing, since it cannot stretch to all three.

The issues surrounding goat production go far beyond the morals of meat eating or animal welfare. It has thrown our entire economy out of gear, with our land and water systems collapsing. In a country where millions of people go hungry, 37 per cent of all arable land is being used to grow fodder for animals that are being raised and killed for export. Government statistics show that the increase in meat production, in particular of beef and veal and buffalo, between 1976 and 1994 has been dramatic – a 17-fold increase of beef and veal from 70,000 tonnes to almost 1.3 million tonnes and a nine-fold increase for buffalo from 216,000 tonnes to just over 12 million tonnes. The value of meat exports grew from Rs 615 million in 1980–81 to Rs 7 billion in 1996–97. The cross-breeding of cows with foreign species for milk has opened another route for the increased slaughter of calves. Since these calves are not hardy enough for ploughing or bullock carts, they are sent for slaughter. As if that were not enough we are even exporting soyabean to feed European livestock, who will in turn be murdered for meat.

These kind of figures cry out against any kind of meat production at all, compassionate or otherwise. I see no reason why India should feed the world at the expense of her own land, her water, her people, her hunger. What are we killing ourselves for? Countries that do not want to destroy their own lands and waters are paying us a minuscule amount to squander these invaluable resources. And we are falling into this trap. We are killing ourselves in a slow spiral.

Five years ago, we set up a Wasteland Development Authority whose sole function was to give out money to local NGOs (non-governmental organizations) to restore land denuded by goats. The Ministry gives out Rs 6000 an acre to any organization that will set out to regreen it. So since each goat eats 10 acres before it is killed, it actually ends up costing the government Rs 60,000. But even this is not the end of it. The Rs 6000 does not pay for fencing, so even when the NGO replants laboriously, the goats are soon back and all that money and work is for nothing.

Nor does this money come cheap. We get it at considerable rates of interest from the World Bank and other finance organizations and find ourselves trapped by debt into kowtowing to all sorts of conditions demanded by the donor agency. More loans for regreening also mean less loans for other social welfare schemes like schools and roads and infrastructure. So a developing country now lacks the means to develop. And all this devastation all on account of sending all that meat abroad.

RECLAIMING THE LAND AND WATER

I have two weekly television shows on a national network. The first is 'Heads and Tails' which focuses on specific practices that are cruel to animals and ways to stop them. The second is 'Maneka's Ark' which involves experts looking at a specific issue that involves more animal-friendly ways to live. Together, the two bring in a couple of hundred letters a day from viewers who have turned vegetarian or been prompted to start a shelter and so on. Recently, I had on a group of young people who left college to go to a place called Alwar in Rajasthan. There, they found that the area used to be sustained by a river which had simply dried up and disappeared. After it had gone, all the small farmers and families that it had sustained had turned into labourers and servants, families had broken up under pressure and murder and crime were common.

When these people came to Alwa, they took only one action – they stopped all the goat breeding around what used to be the river. Within four years, the region regenerated itself. The shrubbery became undergrowth, became woods, and the river returned to life. It is now in full spate and is flush with fish, so naturally the government is trying to get it back so that it can give it out to fish contractors and kill everything all over again.

It took these young people one year to band together the 70 villages that are part of this miracle because, initially, no one was prepared to believe that just taking away the goats would bring back the river. But having achieved it, they guard it jealously. Farming has resumed, families have recovered, family size has voluntarily been restricted to two and prosperity has risen. No goats are allowed in the area. Recently, a local landlord brought in a thousand goats for a marriage ceremony, but they were turned back at the borders of what the villagers now see as their little kingdom. They did not care whether the goats were coming in to be kept or be killed, they just would not have them there even for a day. Should you visit India, after you have done the Taj, be sure to take in this modern miracle in Alwar as well.

A COWDUNG ECONOMY

If India stopped growing and exporting goats and buffaloes, we could regenerate our land and water, we could avert famine, we could increase rainfall, we could find more of everything for everybody, we could transform ourselves by the one single act of ending meat exports. Unfortunately, we are trying very hard now to change from a controlled economy which was run into the ground by the public sector, into a completely capitalistic state, which would take us even faster to the grave. This is liberalism, which means anybody can come in. Except that the companies that are coming in to supposedly help strengthen India are not making roads or schools or any key projects. Instead, Kentucky Fried Chicken, McDonald's and Pizza Hut are the first fruits of liberalization, followed by Bacardi, Jack Daniels and all the Marlboro men.

I do not think that India can be seen in terms of capitalists or communists, it is instead a cowdung economy. If you take the cow or its cowdung away, we are done for. We will die as a people. I say this knowing full well the scorn and ridicule that our dependence on cattle attracts from the developed world. There are clever jokes about Indians who regard cows as their mothers and so on. But scoff at it or not, it is true – we owe a major part of our survival to our cattle.

In old India, Christine Townend (Chapter 4) described how the Panchayat system used to ensure that when cows retired they went into a *gaushala* or cow shelter. Every village maintained one and every home contributed a little grain for it. Now, with land at a premium, the first thing that has gone are the animal support systems. There are no more grazing areas for cattle because they are all being sold off for buildings. And with no place to go, the slaughterhouse becomes their ultimate destination.

The Indian reverence for the cow may seem like sentimentality but it is born from solid practical considerations. In India, all essentials are transported by road. The entire rural economy moves on four legs. A government study in 1978 estimated that draught animals (cows and bullocks) saved us Rs 78,000 crores (1 crore = 10 million) on fuel and transport costs. Since fuel is imported, we are saving valuable foreign exchange. This reasonable cost of transport also keeps food prices within check. Food is cheap because it comes cheaply. In the absence of draught animals, we would be using even more trucks in a country that does not have the infrastructure to support them. As it is, our roads are clogged with traffic. In Dehli alone, over 10,000 trucks pour into the city daily with food and supplies. A majority of these trucks are old and highly polluting. The increase in them manifests itself in higher pollution levels and pollution-related diseases. Studies reveal that traffic policemen frequently develop TB and cancer. All because draught animals are being replaced on the road and taken off and sent to the butchers. For the same reason, my food is getting more expensive. I pay more money for my food now than last year because the animals that used to carry it are gone to the Middle East.

MODERN SLAUGHTERHOUSES – MASS DISASTER

India's first modern slaughterhouse is called Al Kabir and there are more on the way. It sparked very violent protests from local people, citizens' groups, animal welfare organizations – but nobody cared because nobody really understood the economic fallout then. I brought out a series of articles to explain the issue and this elicited 88,000 responses to the government demanding that the slaughterhouse be closed down. In England, that might have been a significant number, but in India it is not enough to force any action, so the slaughterhouse continues till today.

Here is where its dangers lie. The small roadside butchers shops are dreadfully cruel but the damage is limited to only a few animals, for that is all they can kill. A modern slaughterhouse, on the other hand, is able to kill huge numbers because of the way it operates.

In one slaughterhouse, buffaloes are brought in and lined up. A steel chain is slipped over the hind leg by which it is then winched up. Naturally the leg breaks with the weight. The animal is put on a conveyor belt. Then boiling water is poured over it to sterilize the skin. The buffalo is still alive. Now its throat is cut slightly so that the blood can drip out and be collected for medicines – our iron tonics come from slaughterhouse blood. The animal is still alive. Then a steel rod is used to puncture a hole in the stomach to let out the gas so that the skin can be ripped off.

While the skin is being ripped off, the animal finally dies, almost one and a half hours from when the steel ring was fastened around its leg. This is considered 'modern' in India. It happens to several hundred animals a day and will be happening to more as the government moves to open more such torture chambers.

Al Kabir is the first totally export-orientated slaughterhouse. It is based in Andhra Pradesh, near the capital city, Hyderabad. It has had widespread effects – increasing the scarcity and the price of milk and milk products, depletion of natural organic manure, non-availability of bullocks for ploughing the land. In India, there are mainly small farmers since land holdings are limited to 12 acres per farm. This makes buying tractors difficult, so farmers join cooperatives and lease tractors. Ploughing becomes more expensive, it becomes less flexible, it is dependent upon tractor availability, fuel availability and so on. Cattle poaching is a new crime in India and a direct result of cattle depletion due to meat export.

All animal-product industries are suffering. Shoemakers, dairy farmers, wool manufacturers, local butchers are all in trouble due to the advent of the export-orientated slaughterhouse.

After Al Kabir, the prices of milk in the area are the highest in state. And there is no dung available locally. As a result, 300,000 women who used to collect and sell dung for a living are out of a job. The dung used to be used as cooking fuel. Instead, now people cut trees for fuel. As a result, the trees in the area have reduced considerably. As a result, the weather is hotter and the rain less. As a result, it is harder to grow crops. Again, dung used to be used as organic manure. Now, in its absence, there is a large demand for chemical fertilizers, in fact this state uses more chemical fertilizers than the rest of the country put together. The villagers have lost Rs 910 crores just in the money paid out for chemical fertilizers.

In India, animal farming is not a question of compassion or not. Meat farming in any form is bad for our economy. When we talk about 'compassionate animal farming' that is not possible in India. We have a different farming system in India that cannot be adapted to this. We cannot be reformed.

Three years ago I filed a case against a slaughterhouse in Delhi. Part of the petition was a film that showed how the animals were killed.

With a licence for 800, the slaughterhouse was killing 30,000 a day. The judge fainted when he saw the tape and I won my case.

This became the first slaughterhouse to be shut down. After reopening, it has become the only slaughterhouse to function under supervision and the threat of being closed down again if it violates too many provisions. There are over 50,000 slaughterhouses in India, small ones, big ones, medium ones and now these modern giants. These factories of death are, in turn, going to lead to factory farming, which will prove another source of evil and the subject matter of another discourse.

For now, let me end on a more hopeful note. I am part of a government that is headed by a party that has for the last 50 years proclaimed, as a part of its manifesto, a belief in vegetarianism and non-violence. Unfortunately, it has come to power in a coalition of very 'modern' other parties that oppose this. However there are four or five members of the government who, along with me, are lobbying to stop meat export. Now, if the government abides by its convictions, I feel sure that we can expect some changes shortly.

Part IV
Human Well-being

The meat business pushes high meat and animal product consumption as the norm. But this affects human well-being becausc of its effects on our health and the command over resources meat eaters exercise. What we eat can make our lives more or less agreeable, as Chapter 11 demonstrates. For all to eat well requires a broad range of actions in many different areas that go to make up a sustainable food policy – as is discussed in Chapter 12.

11 Feeding the World a Healthy Diet

Geoffrey Cannon

The diets typically consumed in industrialized countries are an important cause of a number of chronic diseases.[1] The diseases most often identified as being causally related to inappropriate diets include tooth and dental diseases, constipation and some more serious gut disorders, adult-onset diabetes, high blood pressure and stroke, and coronary heart disease.[2]

In addition, reviews and assessments of scientific research now show that what can loosely be described as 'industrialized diets' are also an important cause of a number of common cancers.

The first reports relying on modern scientific research to reach this conclusion were published in the early 1980s. One was compiled by the British researchers Professor Sir Richard Doll and Professor Richard Peto;[3] the other was the responsibility of an expert panel convened by the US National Academy of Sciences.[4] Both these reports had a US context and so did not receive as much attention in Europe as otherwise they might.

The New Consensus on Diet and Cancer

The current agenda is set by the expert report 'Food, Nutrition and the Prevention of Cancer: a global perspective', published by the WCRF in late 1997.[5] I was director of the project that produced this report and was also the report's chief editor (see Box 11.1). The report found that food and nutrition certainly or probably modify the risk of over a dozen relatively common cancers, including those of the mouth and throat, lung, stomach, pancreas, liver, colon and rectum, and breast.

Box 11.1 The WCRF Report on Diet and Cancer

The American Institute for Cancer Research (AICR), the original parent organization of WCRF, was founded in 1982. WCRF and AICR are unique, in that they are the only national charities in the UK and the US dedicated solely to the prevention of cancer by means of food and nutrition. The foundation of AICR coincided with publication of the National Academy of Sciences (NAS) report[6] and AICR's original dietary guidelines were based on those of the NAS report. By the 1990s it was time for a new report, which was commissioned by WCRF, with AICR, in 1992.

The task was necessarily ambitious. Fourteen scientists from all round the world were formed into a panel, and asked to review and assess the totality of the mainstream scientific literature. Currently, over 500 papers on food, nutrition and cancer are being published in peer-reviewed scientific journals every year, and the panel eventually took over 4500 papers into account. The task was also daunting, because, whereas coronary heart disease is basically one disease entity, from the point of view of cause and prevention cancers are separate entities, sometimes closely related, sometimes not. For example, while dietary factors affecting the risk of stomach cancer, colorectal cancer and breast cancer are in some ways similar, in other ways they are strikingly different. And regular physical activity is certainly protective against colon cancer and probably against breast cancer, but seems to be irrelevant in the case of stomach cancer. Conversely, diets high in salt and salted food probably increase the risk of stomach cancer, but salt seems to have nothing to do with colon or breast cancer.

Nevertheless, the panel agreed that its recommendations should be designed to prevent cancer as a whole. Devising separate dietary recommendations for up to 18 cancer sites would not only be mind-boggling but obviously self-defeating for anybody concerned with clear public health messages.

The panel responsible for the report and its recommendations included Professor Philip James, well known in the UK as the original architect of the Food Standards Agency, and Professor Tony McMichael of the Department of Epidemiology and Public Health at the London School of Hygiene and Tropical Medicine. The panel was chaired by Professor John Potter, an Australian now based in the US. Other panel members came from Italy, India, Japan, China, Mexico and the USA.

The panel was supported in its work by over 100 contributors, advisors and reviewers, and also by official observers from the three relevant UN agencies, the WHO, the FAO and the International Agency for Research on Cancer (IARC). A secretariat, of which I was the head, supported the panel in drafting and redrafting sections and chapters of the report which, in some cases, were reworked ten or more times. The final report is 670 pages in length.

On publication, the report was launched at scientific conferences or special meetings in London, New York, Brussels, India, Africa and Latin

America. It is now generally accepted as integrating the story on diet and the prevention of cancer with diet and the prevention of coronary heart disease and other major chronic diseases. A Chinese language edition will be launched in Beijing in 1999 and a Spanish language edition is due to be published by the Pan American Health Organization.

For a copy of the WCRF's expert report, *Food, Nutrition and the Prevention of Cancer: a global perspective* or its summary, please contact WCRF at 105 Park Street, London W1Y 3FB. The full report is priced £25; and the summary £2.

The report breaks new ground in a number of ways. It emphasises the importance of appropriate foods, drinks, and diets as a whole as protecting against cancer. In particular, it recommends individuals choose predominantly plant-based diets rich in a variety of vegetables and fruits, pulses (legumes) and minimally processed, starchy staple foods. It also emphasises the importance of lifestyle factors closely associated with diet, in particular body weight and physical activity. In the UK, a further report published in 1998 for the Department of Health by the Committee on Medical Aspects of Food Policy,[7] has reached similar conclusions, as have statements published by the Europe Against Cancer programme,[8] a report commissioned by the French government;[9] and reports published in the US by the National Cancer Institute[10] and the American Cancer Society.[11]

It is now safe to say that many cancers, particularly those of epithelial tissues, are as closely causally associated with diet as is coronary heart disease. Around two thirds to three quarters of all causes of stomach cancer and colorectal cancer are estimated to be preventable by appropriate diets and associated lifestyles. Perhaps one third to one half of all cases of breast cancer are preventable by similar means, although there is good evidence that for this cancer, prevention is mostly effective only early in life, before puberty. As one more example, although of course the overwhelming cause of lung cancer is smoking and other use of tobacco, diets high in vegetables and fruits are certainly protective against lung cancer.

Prevalence of Cancer

Cancer is a common killer disease everywhere in the world. Estimates published by the World Health Organization (WHO) in 1997 showed that throughout the world, every year, there are currently around 10 million new cases of cancer identified (Tables 11.1 and 11.2).

Table 11.1 ***Estimated Numbers of New Cases and Deaths from Cancer Worldwide 1996: Men***

Cancer site	*New cases (thousands)*	*Per cent of total*	*Deaths (thousands)*	*Per cent of total*
Lung	988	18.6	878	22.4
Stomach	634	11.9	518	13.2
Colon, rectum	445	8.4	257	6.6
Prostate	400	7.5	204	5.2
Mouth and pharynx	384	7.2	237	6.1
Liver	374	7.1	370	9.4
Oesophagus	320	6.1	305	7.8
Bladder	236	4.4	107	2.7
Other	1,531	28.8	1,043	26.6
Total	5,312	100.0	3,919	100.0

Source: WHO (1997) *The World Health Report 1997*, World Health Organization. Geneva

Table 11.2 ***Estimated Numbers of New Cases and Deaths from Cancer Worldwide 1996: Women***

Cancer site	*New cases (thousands)*	*Per cent of total*	*Deaths (thousands)*	*Per cent of total*
Breast	910	18.2	390	12.2
Cervix	524	10.5	241	7.6
Colon, rectum	431	8.6	253	7.9
Stomach	379	7.6	317	9.9
Lung	333	6.7	282	8.8
Mouth and pharynx	192	3.8	129	4.1
Ovary	191	3.8	125	3.9
Endometrium	172	3.4	68	2.1
Other	1,874	37.4	1,387	43.5
Total	5,006	100.0	3,192	100.0

Source: WHO (1997) *The World Health Report 1997*, World Health Organization. Geneva

In men, the most common cancer is of the lung, followed by stomach, colorectal and prostate cancer. In women, the most common cancer is of the breast, followed by cervical, colorectal and stomach cancer. Diet modifies the risk of most of these cancers. Cancer incidence is increasing both absolutely and relatively (Table 11.3). To some extent, the absolute increase is a function of an increasing and ageing human population. The relative increase is more striking. Between 1985 and

2015, the proportion of deaths from cancer as a percentage of deaths from all causes is projected to double in Africa, Latin America and Asia. This indicates a profoundly important shift in the relative importance of different types of disease worldwide. In the economically developing world, endemic deficiency and infectious diseases will remain common. But to an increasing extent, people in Africa, Latin America and Asia, especially in urban areas, are now suffering and dying from the same type of chronic diseases that afflict us.[12]

Table 11.3 *Major Causes of Death in the World: 1985 and 2015*

Cause of Death (As percentage of all deaths)	*Economically Developing Countries*		*Economically Developed Countries*	
	1985 (%)	*2015 (%)*	*1985 (%)*	*2015 (%)*
Infection	36	19	9	7
Cancer	7	14	18	18
Circulation	19	35	50	53
Pregnancy, perinatal	9	6	1	1
Injuries	8	7	6	5
Other	21	19	15	16
		Millions		
Total deaths from all causes	37.90	47.80	12.00	14.50
Total deaths from cancer	2.65	6.69	2.16	2.61

Source: Table 9.2.4 in WCRF/AICR (1997) *Food, Nutrition and the Prevention of Cancer: a global perspective*, American Institute for Cancer Research, Washington, DC, p 553

DIETS AND FOOD SUPPLIES

Evidence that the diets typically eaten in industrialized countries such as the UK and the US are a major cause of coronary heart disease and other diseases of the circulation system has been accepted as amounting to proof beyond reasonable doubt for 20 or 30 years now. Public health policies, agreed by international agencies such as WHO, and national governments, throughout the industrialized world, now usually include recommendations to consume less saturated fat and fat generally (and usually also less sugar and salt) and, more positively, to consume a lot more vegetables, fruits and starchy staple foods, preferably wholegrain.[13]

It is sometimes thought that what we eat is just a matter of individual choice. This of course is not so. Individual diets are largely a function of national and regional food supplies.[14] It is not always appreciated that the food supplies of countries that are now completely industrialized changed radically as a result of industrialization. Take the UK as an example. Beginning in the 17th century, the English parliament sanctioned the enclosure of previously common land, partly as an instrument of power and partly as a means of increased production of crops and animals. This process was repeated in Scotland and Ireland for the additional reason of subjugating rebellious native people. Beginning in the 18th century, with mechanization and industrialization, the process by which rural people were driven off the land into new and vastly increasing cities was accelerated. By the middle of the 19th century, England had become mainly an urban country, and the peasantry had largely disappeared; not by choice, but as a consequence of industrialization. As a further consequence the national food supply changed radically.

This process accelerated further between the mid 19th century and the early 20th century, as a result of a series of inventions and developments that had the general effect of degrading the food supply further. These included the introduction of steel roller mills, making the mass production of white flour and bread possible; the lifting of the tax on sugar, which made sugar the most profitable edible food commodity; and hydrogenation, the process by which oils are turned into saturated fats, that founded the fortunes of the margarine industry. Also, the industrialization of death, first devised in the stockyards of Chicago, involving vast disassembly lines and the use of the new intercontinental railways, made meat and meat products for the first time an everyday commodity for poor as well as for rich people in the US, and then later in Europe.[15]

What we typically eat in the UK is not merely a reflection of individual choice, but has generally been determined by impersonal factors which, in the case of the UK, are generally historical in origin. Different countries have different stories. In the US, Canada, Australia, New Zealand and South Africa, immigrants from Europe and elsewhere displaced native populations. Generally speaking, countries in continental Europe have suffered the ill-effects of industrialization rather less dramatically than England, but the general point remains true, that the process of industrialization and urbanization creates food supplies and, therefore, diets that are very different from those produced and consumed traditionally.

Food manufacturers of course do not devise foods in order to poison the population. But, as a general rule, food supplies made up of more and more processed foods do increase the risk of chronic diseases, and the more processed food is, the more this tends to be true. White bread is relatively short of fibre, essential fats and many micro nutrients. Sugar, which makes otherwise disgusting or insipid ingredients palatable, itself has no nutritional value other than as a source of calories, obviously unhelpful for sedentary populations. Hydrogenated fats of vegetable and animal origin are found in most manufactured foods. Altogether, these create food supplies that are increasingly lifeless. Although it is not a very scientific thing to say, the more that food supplies are made up of fresh foods, the lower the risk of chronic diseases, for a constellation of reasons.

New epidemics

The panel responsible for the WCRF report was particularly concerned with what is now a global phenomenon, known as the demographic–nutritional–epidemiological transitions (Figure 11.1). This phenome-

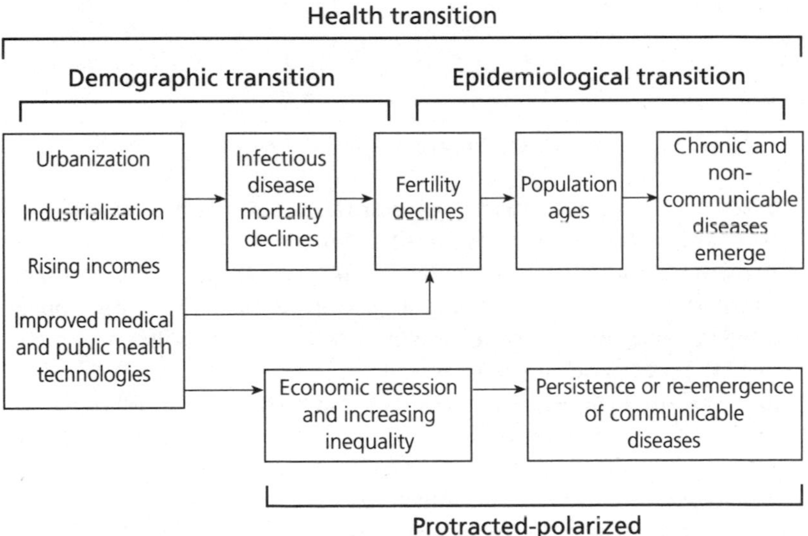

Source: Taken from Figure 9.2.1 in WCRF/AICR (1997) *Food, Nutrition and the Prevention of Cancer: a global perspective*, American Institute for Cancer Research, Washington, DC

Figure 11.1 *The Transition from Deficiency Diseases and Infection to Chronic Disease Patterns*

non, noted in particular by Professor Barry Popkin of the University of North Carolina at Chapel Hill,[16] is now repeating in Africa, Latin America and Asia, the pattern first set in England hundreds of years ago. Throughout the economically developing world, rural populations are moving from the countryside into cities, typically because they have no choice (Table 11.4). As a result, the nature of the food they eat changes dramatically.

Table 11.4 *Percentage of Population Living in Urban Areas*

Region	*1995*	*2025*
World	45.2	61.1
Economically developing regions (average)	37.0	57.0
Africa	34.4	53.8
Asia	34.6	54.8
Latin America and the Caribbean	74.2	84.7
Economically developed regions (average)	74.7	84.0
Australia/New Zealand	84.9	89.1
Japan	77.6	84.9
Northern America	76.3	84.4
Europe	73.6	83.2

Source: UN World Urbanization Prospects: The 1994 Revision

Consequently, the patterns of disease of these new urban populations also change dramatically. Specifically, chronic diseases become much more common. These include, in rough order of emergence: constipation, overweight and obesity, high blood pressure, adult onset diabetes, lung cancer, colon cancer, stroke, coronary heart disease, prostate cancer and breast cancer. Thus, these populations begin to suffer and die from the same diseases that afflict us, but often earlier in life than us. For nations in the economically developing world, this phenomenon is a catastrophe. First, because these countries still suffer from endemic deficiency and infectious diseases, and also from some endemic cancers. So the disease burden of these countries, already much heavier than that of rich countries, becomes intolerable. With a few exceptions, even those countries in Africa, Latin America and Asia that are comparatively rich in resources are much less prosperous than economically developed countries. They simply do not have the money to screen for, treat and palliate major killer chronic diseases. This is largely because the world's richest nations still enjoy the legacy of exploitation of former colonies, or else still actively exploit other countries by economic if not political and military methods.

HEALTHY FOOD SUPPLIES AND DIETS

The scale of the 670-page WCRF report reflects a vast amount of scientific evidence, in many cases amounting to proof beyond reasonable doubt, that diets together with associated lifestyles crucially affect cancer risk. The panel responsible for the report also took care to point out that its recommendations designed to reduce the risk of cancer by, at their estimate, between 30 to 40 per cent, also have the effect of reducing the risk of other major diet-related chronic diseases, notably coronary heart disease. The report recommends the policy framework for global cancer prevention which is reproduced here in Box 11.2. The panel also, in its policy recommendations, devised a more detailed framework for policy initiatives that can be carried out by international agencies and national governments (see Box 11.3).

WHAT YOU CAN DO

In common with most expert reports published in the last five years or so, and also reflecting the scientific evidence, the panel's recommendations are based on foods and drinks rather than on dietary constituents such as fat or vitamin C. Food-based dietary guidelines have the additional advantage of being easy to understand and to follow.

The first recommendation made by the panel addressed to individuals is:

> *1 Choose predominantly plant-based diets rich in a variety of vegetables and fruits, pulses (legumes) and minimally processed starchy staple foods.*

The recommendation to choose plant-based diets is original. It is a very broad conclusion based on the totality of the scientific evidence. The panel was aware that this recommendation, when implemented, also implies sustainable food and agricultural systems.

BOX 11.2 A POLICY FRAMEWORK FOR CANCER PREVENTION

The panel invites international agencies and national governments, industry, the medical and health professions, consumer and public interest groups, the media, and all other organizations worldwide, to accept and act on the following judgements, based on current scientific knowledge.

The scale of benefit:

- Cancer is mostly a preventable disease. The chief causes of cancer are use of tobacco, and inappropriate diets.
- Between 30 per cent and 40 per cent of all cases of cancer are preventable by feasible and appropriate diets and related factors.
- On a global basis and at current rates, this means that appropriate diets may prevent 3–4 million cases of cancer every year.
- Diets containing substantial and varied amounts of vegetables and fruits will prevent 20 per cent or more of all cases of cancer.
- Keeping alcohol intake within recommended limits will prevent up to 20 per cent of cases of cancers of aerodigestive tract, the colon and rectum, and breast.
- Cancers of the stomach and colon and rectum are mostly preventable by appropriate diets and related factors.
- A feasible intermediate target for the dietary prevention of cancer is the reduction of global incidence by 10–20 per cent within 10–25 years.

The need for prevention:

- Prevention of cancer by dietary and associated means, and by prevention and cessation of smoking, are the most effective approaches.
- Prevention benefits not only individuals, but also families, communities and national economies.
- Prevention is the only sensible approach to cancer in the developing world; on a population basis, treatment and palliation of cancer are economically not feasible.

Taken from WCRF/AICR (1997) *Food, Nutrition and the Prevention of Cancer: a global perspective*, American Institute for Cancer Research, Washington, DC, p 548

2 ***Avoid being underweight or overweight and limit weight gain during adulthood to less than 5kg (11 pounds).***

3 ***If occupational activity is low or moderate, take an hour's brisk walk or similar exercise daily, and also exercise vigorously for a total of at least one hour in a week.***

These second and third recommendations naturally go together. Physical inactivity is of course one of the main reasons why people become overweight and obese; and there is convincing evidence that people who are physically active and whose body mass is within a recommended range, are much less likely to suffer from a number of common cancers. This recommendation is equally important as a means to prevent coronary heart disease.

> 4 *Eat 400–800 grammes (15–30 ounces) or five or more portions (servings) a day of a variety of vegetables and fruits all year round.*

This 'Five plus a day' recommendation for vegetables and fruits is now fairly well known. This, too, applies to coronary heart disease as well as to cancer. Average consumption of vegetables and fruits in the UK and other northern European countries is around three portions a day and so this recommendation represents a big increase for almost everybody.

> 5 *Eat 600–800 grammes (20–30 ounces) or more than seven portions (servings) a day of a variety of cereals (grains), pulses (legumes), roots, tubers and plantains. Prefer minimally processed foods. Limit consumption of refined sugar.*

The evidence that starchy staple foods (preferably wholegrains) eaten in abundance protect against coronary heart disease is stronger than the direct evidence that these foods protect against cancer. However, they are obviously preferable to diets that are relatively fatty or sugary.

> 6 *Alcohol consumption is not recommended. If consumed at all, limit alcoholic drinks to less than two drinks a day for men and one for women.*

It is now commonly thought that substantial consumption of alcohol protects against coronary heart disease. This notion, encouraged by the trade, is not true. There is good evidence that one or, at the most, two drinks of alcohol a day protect against coronary heart disease, but there is no good evidence that consumption above this level is protective. The evidence relating to a number of common cancers taken together means that the only recommendation that can be made solely with cancer in mind, is not to drink alcohol at all. However, non-

BOX 11.3 INTERNATIONAL AGENCIES AND NATIONAL GOVERNMENTS: SUGGESTIONS FOR POLICY INITIATIVES DESIGNED TO PREVENT CANCER

The panel invites national governments, supported by international agencies, to incorporate the prevention of cancer into relevant policies and programmes by initiatives such as those suggested here. These can be adapted to suit economic, political, economic and local circumstances. Such policies and programmes generally require government initiation and continued support. The suggestions are not comprehensive; they are designed to stimulate programmes of action that generally require collaboration between all the agents of change identified in this chapter.

General

- Incorporate a strategic public health dimension into all relevant international and national political economic policies.
- Identify the prevention of cancer as a key objective at global, regional, international, national, and local levels.
- Compile, update and disseminate case studies of international, national and local policies and experience relevant to the prevention of cancer.
- Sustain national food agencies or similar bodies with access to all ministries whose policies affect food, nutrition and public health.
- Ensure equitable representation on such agencies from representatives of industry and health, medical, consumer and public interest groups.
- Enable such agencies to assess food supply systems 'from plough to plate' to ensure typical diets protect against disease.

Legislation

- Examine, audit and revise existing and proposed legislation relevant to the prevention of diet-related diseases.
- Appoint ministers and senior officials with responsibility for public health and, in particular, for the prevention of cancer and chronic diseases.

Economic

- Assess the general effect of political and economic policies on public health and, where appropriate, use public sector investment to protect public health.
- Estimate the economic impact of the burden of cancer, allowing for projected increases in population and changes in distribution.
- Maintain and increase taxes on alcohol. Countries where alcohol is discouraged by a variety of means should maintain such policies.
- Ensure that adequate funding is available to ensure that publicly financed institutional catering can meet adequate nutritional standards.

Development

- Include the cost of policies to prevent cancer and other chronic diseases within international and national development plans.
- Address, as an integral part of public health strategies, the impact of urbanization and industrialization on patterns and incidence of cancer.
- Audit the impact of the globalization of food trade on the projected incidence of cancer, internationally and nationally.
- Ensure that urban development and regeneration provides for and protects green belts, street markets, and inner city food-shopping facilities.

Agriculture

- Emphasize the production of foods of plant origin: vegetables and fruits, and cereals, tubers, roots and pulses for human consumption.
- Encourage sustainable and appropriate agriculture to produce foods important in diets that prevent chronic diseases, including cancer.
- Assess the land, water, energy, and all other resource needs for sustainable agriculture systems most likely to produce plant food.
- Reconsider the effects both of traditional and of modern agriculture on the nutritional adequacy and quality of the foods produced.
- Compare the relative benefits of cash crops grown for immediate economic return, and agriculture whose products protect public health.
- Encourage the market for foods of plant origin, especially vegetables and fruits, including, if appropriate, by price-support systems.
- Reconsider price market support systems that create artificial markets for foods from land animals, in particular fatty foods.
- Review the ecological, public health and long-term economic impact of the rearing of land animals.

Health

- Integrate analyses of long-term public health outcomes into strategies designed to sustain international and national economic growth.
- Recognize the economic benefits of primary health care, including prevention and screening for cancer.
- Support cancer prevention proportionately to prevention and treatment of deficiency and infectious diseases, and other chronic diseases.
- Integrate cancer prevention into programmes to prevent deficiency, infectious, and other chronic diseases.
- Integrate cancer prevention into programmes to ensure food security and safety.
- Give increasing public funding to prevention of cancer up to appropriate levels in the general context of promotion of public health.
- Reconsider the proportions of public funding given to the treatment of cancer, compared with funds designed to screen for and prevent cancer.

- Set targets for prevention of cancer over specified periods of time in a general framework of disease prevention.
- Establish programmes designed to achieve the dietary goals recommended in this report, in the short term and the longer term.
- Ensure that such programmes are effective, by means of multi-disciplinary, multi-agency networks and alliances at all levels.

Education

- Ensure that school curricula include teaching on food, nutrition and health, and on the importance of active living.
- Ensure that school children at all stages have proper access at school to healthy meals, and to recreation and sports facilities.

Transport

- Encourage local food production and distribution systems that minimize long-distance transportation of food and drink.
- Develop transport systems that encourage walking and cycling, and facilities that encourage physical activity throughout life.
- Allocate a proportion of transportation budgets for the development of bicycle and pedestrian facilities, notably in urban areas.

Taken from WCRF/AICR (1997) *Food, Nutrition and the Prevention of Cancer: a global perspective*, American Institute for Cancer Research, Washington, DC, p 570

smokers who consume an otherwise healthy diet are unlikely to be at greater risk of any cancer if they consume at most one or two drinks a day.

> 7 ***If eaten at all, limit intake of red meat to less than 80 grammes (3 ounces) daily. It is preferable to choose fish, poultry or meat from non-domesticated animals in place of red meat.***

Above a certain level of intake, red meat probably increases the risk of colon cancer and possibly increases the risk of other cancers. The evidence relates to red meat only, and not to fish or to poultry or to meat from non-domesticated animals. A maximum of 80 grammes a day of red meat is somewhat less than current average UK consumption. While there is no evidence that a vegetarian diet is preferable to a healthy diet that also includes small amounts of red meat, there is no need to eat meat at all.

8 *Limit consumption of fatty foods, particularly those of animal origin. Choose modest amounts of appropriate vegetable oils.*

The evidence that fat as such is a direct cause of cancer is now considered to be rather less strong than it was ten or more years ago. However, fatty diets certainly increase the risk of overweight and obesity, which increase cancer risk.

9 *Limit consumption of salted foods and use of cooking and table salt. Use herbs and spices to season foods.*

This recommendation relates to the risk of stomach cancer, incidence of which is rapidly decreasing throughout the world. It is generally agreed that a major reason for this is replacement of salting by refrigeration. The panel emphasises that salty diets increase stomach cancer risk as well as diets containing substantial amounts of salted foods.

10 *Do not eat foods which, as a result of prolonged storage at ambient temperatures, are liable to contamination with mycotoxins.*

This recommendation mainly applies to tropical countries where incidence of primary level cancer, probably caused by mycotoxins, is high.

11 *Use refrigeration and other appropriate methods to preserve perishable food as purchased and at home.*

This recommendation, while universal, is especially important in countries where the use of industrial and domestic refrigeration is still relatively uncommon.

12 *When levels of additives, contaminants and other residues are properly regulated, their presence in food and drink is not known to be harmful. However, unregulated or improper use can be a health hazard, and this applies particularly in economically developing countries.*

There is a high level of public anxiety about additives and contaminants in food. However, in countries like the UK, there is no good evidence from studies of humans that chemical residues in food are a significant factor in the risk of any cancer. Certainly, in the case of pesticides, any conceivable cancer risk is outweighed by the benefits of consuming plenty of vegetables and fruits which may contain detectable trace amounts of chemical residues.

> 13 *Do not eat charred food. For meat and fish eaters, avoid burning of meat juices. Consume the following only occasionally: meat and fish grilled (broiled) in direct flame; cured and smoked meats.*

General good advice is to avoid burning. This applies to sunburn, and to consumption of extremely hot or abrasive food and drink. It also applies to food which is itself burned or cooked in direct flame.

> 14 *For those who follow the recommendations presented here, dietary supplements are probably unnecessary, and possibly unhelpful, for reducing cancer risk.*

In certain contexts, dietary supplements are valuable. However, there is, as yet, no good evidence from studies of humans that any combination of dietary supplements protect against cancer. Indeed, some recent large trials suggest that synthetic supplements may even increase cancer risk.

> 15 *Do not smoke or chew tobacco.*

The brief of the expert panel did not include any review or assessment of the literature on smoking or other use of tobacco. However, in common with other expert panels, the scientists responsible for the report decided to warn against tobacco use, because of the overwhelming evidence that it increases risk not only of lung cancer, but also cancers of the oral cavity and respiratory tract.

NOTES

1 WHO (1990) *Diet, nutrition and the prevention of chronic diseases Report of a WHO study group*. Geneva (Technical report series 797)
2 Cannon G (1992) *Food and health: the experts agree*, Consumers' Association, London
3 Doll R and Peto R (1981) 'The Causes of Cancer: quantitative estimates of avoidable risk of cancer in the United States today', JNCI. 66, pp 1191–1308
4 NAS (1982) *Diet, nutrition and cancer*, National Research Council, National Academy of Sciences, National Academy Press, Washington, DC
5 WCRF/AICR (1997) *Food, Nutrition and the Prevention of Cancer: a global perspective*, American Institute for Cancer Research, Washington, DC
6 See ref 4
7 COMA (1998) *Nutritional aspects of the development of cancer*, Committee on Medical Aspects of Food and Nutrition Policy, Department of Health, UK
8 Boyle P, Veronesi M, Tubiana FE, et al (1995) European School of Oncology Report to the European Commission for the 'Europe Against Cancer Programme', European Code Against Cancer, *Eur J Cancer*, Vol 31A (9), pp 1395–1405
9 CNERNA/CNRS (1996) *Alimentation et Cancer: évaluation des données scientifiques*, Riboli E and Collet-Ribbing (Eds) Centre national d'études et de recommendations sur la nutrition et l'alimentation (CNERNA) and Centre national de la recherche scientifique (CNRS), Paris
10 NCI (1994) *5 a day for better health Program*, Program Guidebook, National Cancer Institute: Bethesda, Maryland
11 ACS (1996) 'Guidelines on diet, nutrition and cancer prevention: reducing the risk of cancer with healthy food choices and physical activity' American Cancer Society, Washington, DC, CA *Cancer J Clin*, 46, pp 325–341
12 Shetty P S and McPherson K (1997) 'Diet, Nutrition and Chronic Disease: Lessons from Contrasting Worlds' Shetty PS and McPherson K (Eds) Proceedings of the Sixth London School of Hygiene Public Health forum
13 See ref 1
14 See ref 2
15 See ref 2
16 Popkin B M (1994) 'The nutrition transition in low-income countries: an emerging crisis' *Nutr Rev* 52, pp 285–298

12 Towards a Sustainable Food Policy

Tim Lang

Today, if we want a food policy to support more sustainable and humane practices we must consider several key themes:

1. To change the food system requires building alliances – holy and unholy alliances, short and long term.
2. We have to get our message from the margins to the centre. Some groups are doing that better than others. Individually, we sometimes succeed, we sometimes fail. Collectively, we need to pull together at the same time and in the same way with the same message – across the environment, health, social justice, animal welfare, consumers, and public health arenas. In welding those messages together we will build a strong alliance. Strides have been made in forging this alliance over recent years, but, frankly, this needs to go further, faster and to be more international.
3. Public policy in this area has to accept (and confront) a very complex food system. Even though simple messages are very important, we must acknowledge that food, farming and animal welfare issues are immensely complex. It is one thing to reduce our messages to 'sound bites'; it is another to believe that the sound bite covers it all.
4. The food policy world, like all sectors, is replete with myths. The hi-tech, agricultural and food world has a myth that everything small is bad. Equally, its opponents have myths as well – that hi-tech is automatically bad, that small is always better, that intensive farmers do not care. We need to be acutely aware that myths convey truths but that we need constantly to review them. So locked are we into positions we do not like that it might be useful

to take time out and reflect on our own myths. Some might need revision and up-dating. Sometimes, I suspect, we might have to admit we are talking emotional hogwash and lack evidence.

5 There are enormous gaps in our understanding, but we know enough about 'alternative' views of food policy to be able to sketch sensible options for public policy makers. Whether we can get them to listen is partly up to us and the strength of our opponents.

I see a number of problems in current food policies. These are:

- Cost externalization: this is the argument that the cost of conventional food and agriculture is not fully reflected in the price of goods. Costs are dumped onto the environment, dumped onto the poor – both here and in the developing world – dumped onto health, dumped onto consumers and dumped onto animals.
- Social division: even though our attention is on animal welfare, let us not forget that the food system operates within a world marked by terrible social divisions. The gap between rich and poor is great.
- Concentration of control: immense power over food is being accumulated in fewer and fewer hands.
- A lengthening of the food chain: food is travelling further and although shop shelves might be full, it is hard to know what is happening before the food gets there.
- A replacement of labour by capital: the 20th century has witnessed extraordinary application of technology: tractors, agrichemicals and now genetic engineering. Who is benefiting?
- Waste: it is not just animals who are wasted, energy is wasted, so are lives and labour.
- Biodiversity: behind the apparent biodiversity on supermarket shelves lies a rapidly shrinking biodiversity on the land.

Before examining these further, let us clarify what we mean by food policy. Food policy is often seen as a 'top down' governmental activity. The World Bank, for example, has argued that 'Food policy encompasses the collective efforts of governments to influence the decision-making environment of food producers, food consumers, and food marketing agents in order to further social objectives.'[1] The Organisation for Economic Co-operation and Development (OECD) suggests something similar. Food policy, it says, is '... a balanced government strategy regarding the food economy, which takes account of the interrelationships within the food sector and between it and the rest of the national and international economy.'[2]

These definitions, and others that focus on health aspects of food policy, are too government focused. They assume that governments have the power or inclination to set food policies. They also ignore the realities of the food economy – which is highly capitalized and concentrated and where the dominant companies are very powerful. Once people lose access to land (or gardens) they are a captive, consuming class. Governments are neither all powerful nor completely powerless, as some of the more extreme proponents of globalization argue, when it comes to setting food policies able to deal with the complexity of the modern food system.

For me, food policy is about the decision-making processes which affect who eats what, when, where and on what conditions. In trying to understand food policies, we try to piece together the different areas of activity which affect the total picture of food production, distribution and consumption. In the UK since World War II, food policy has been dominated by productionism.[3] After a century of external reliance, the 1947 Agriculture Act symbolized a shift to supporting British farming and encouraging production. This policy change enabled the scientific revolution to be more systematically applied, both on and off the farm. Intensification was the theme, getting more from capital, labour and animals. Entry into the CAP changed the financial regime supporting agriculture but the message of intensification stayed the same.

Today, one of the main trends in the food sector is concentration. Everything is getting larger – retail stores, companies, farms or research institutes. People are also travelling further to get their food and, at the same time, the food is travelling further. All of this has a hidden energy cost – which is why many observers now think that this approach to agriculture and food actually cannot be sustained. It uses non-renewable resources brilliantly but not efficiently.

Poverty, overproduction and waste

For consumers – that is us – the modern food system offers immense attractions. Spending less money on food means we can spend more money on clothes, cars, and enjoying ourselves. Alongside this consumerism, however, there is an immense social fragmentation. No longer is there a simple gap between rich and poor in the 'developed' and 'developing' countries. The gap between the rich and poor in the world is such that the middle classes of the UK live closely similar lifestyles to the rich of the developing countries. The world's divisions are moving horizontally, not just vertically. One result shows up in massive inequalities in health between rich and poor.[4] Globally, there is a massive food poverty problem. The United Nations Children's Fund (UNICEF) calculates two billion people of the world's 5.7 billion are affected by food-related ill health due to income deficiencies. Of those, 800 million are undernourished. Even here in Europe, poverty is immense: a staggering 23 million households, or 57 million people out of the EU's 300 million, are below the EU poverty line. Unfortunately, the UK – which has led Europe in being the strongest proponent of the free market and neo-liberal policies – accounts for 30 per cent of all the officially defined poor in Europe.[5]

When I sat on the UK government's Food and Low Income Working Party in the UK it was clear there is a staggering level of food poverty in the UK. About eight million people eat a diet below the WHO recommended nutritional targets due to lack of money. The WHO sets daily targets as:[6]

- a vegetable and fruit intake of at least 450 grammes;
- total and saturated fat intake of no more than 30 per cent and 10 per cent of energy;
- average intake of bread, potatoes and grains to be at least 60 per cent of energy;
- intake of 40 grammes dietary fibre;
- sugar intake no more than 10 per cent of energy;
- (iodized) salt intake no more than 6 grammes.

It also supports breast feeding for children. Scandalously, despite these needs, thousands and thousands of tonnes of fruit and vegetables are destroyed every year in the EU due to 'overproduction' – meaning that people cannot afford to buy them. In 1993–94, the EU spent 390 million ECU to buy and destroy 2.5 billion kilos of fruit and vegetables.[7] Why not give schoolchildren the 980 million kilos of apples or 312 million kilos of oranges destroyed? This is waste on a grand scale – of people's health and good produce – because they cannot afford them, not because they do not need them.

It would be too simple just to dismiss these practices in the EU as the madness of the CAP. The US, for instance, has equally disgusting policies and other ways of dumping unwanted foods on the poor. It has given away surplus dairy fats (which can be bad news for arteries!) for many years as food benefits. We have to recognize that there is a problem of mismatch and overproduction, a tendency to overproduction, alongside a massive problem of underconsumption and inequalities of consumption.

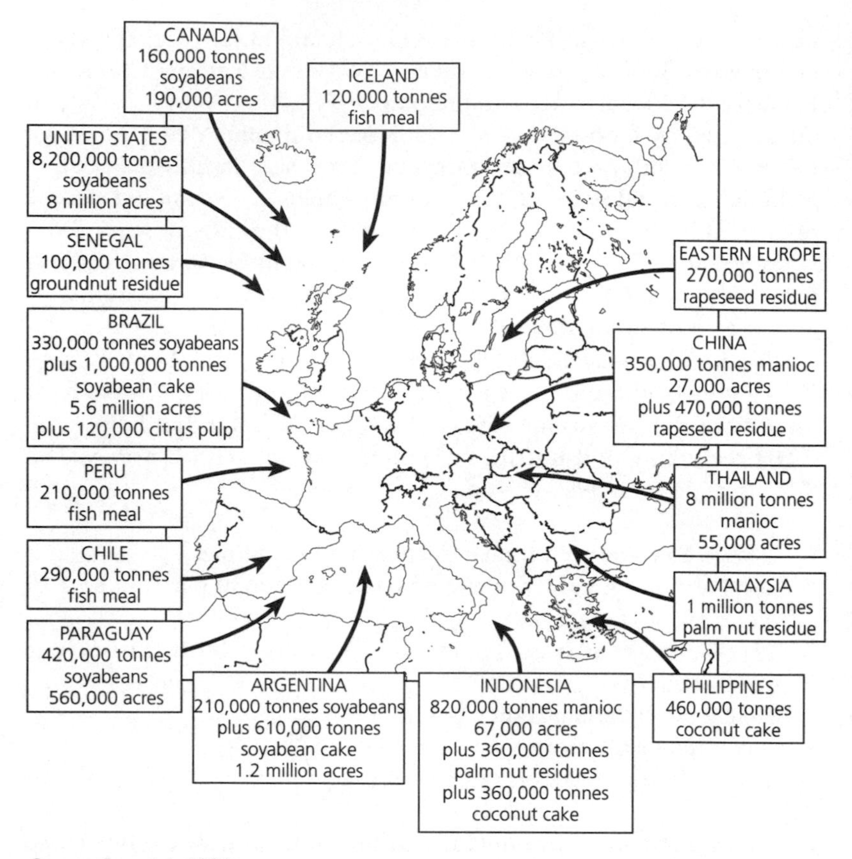

Source: Eurostat, 1993

Figure 12.1 *'Ghost Acres': Main Imports of Animal Feed Ingredients into Europe in 1993 and Selected Acreages*

Neither Europe nor the US is self-sufficient. They draw in food, labour, capital and non-renewable energy in different ways. Figure 12.1 (map) illustrates this for Europe, which draws on the rest of the world to feed its farm animals.[8] In Brazil alone, the equivalent of 5.6 million acres of land is used to grow soya beans for animals in Europe. These 'ghost acres' belie the so-called efficiency of the hi-tech agriculture and food model that Dennis Avery (see Chapter 2) and his colleagues promote. It may be brilliant but it is pretty inefficient. Neither the EU nor the US have efficient production. Rather they have different fiscal structures, different agricultural supports, but they are both hi-tech, intensive agricultural systems. In their study, *Tomorrow's*

World, FoE calculates that the UK was importing 4.1 million hectares of other people's land in 1996.[9]

How we use land is a crucial issue for the future. We need to use less land to produce the same amount of food. If the high population scenario of 11.9 billion people by the year 2020 happens, we are going to need to grow food at about double the intensity. The existing model of agriculture has to incorporate the biotechnology revolution.[10]

Another way of thinking is also making its claim: the ecological approach. This suggests that we need to go back to first principles. What is farming for? If we say it is for feeding people, why, for instance, try to improve animal efficiency if the nutritionists argue it would be good for health (and healthcare) if we ate less? The ecological model suggests that we try to farm food in a way which is good for humans and ecology. Take fruit, for example. The British ought to eat more fresh fruit and vegetables to meet health targets, but we are importing more fruit and vegetables than we export. Our climate and land are good for growing fruit and vegetables yet production is not meeting demand (or health targets). There is no sense in having a balance of payments deficit on fruit and vegetables. In Finland, however, which used to have the rich world's highest rates of heart disease, they have pioneered and redeveloped the growing of berries, which flourish in those northern climes. Why import other fruits, using others' precious landspace, when you could grow it yourself? That is the policy question. The Finns' food policy is not to create a hermetically sealed, old-style protectionist food economy, but to reduce unnecessary food miles (the distance food travels before consumption) and meet health needs.[11] This is the kind of approach I think the UK should also adopt.

Health costs

Modern agriculture externalizes costs onto health. Food is often a major factor in contemporary ill health, certainly in heart disease and food poisonings, but these costs are not reflected in the cost of the UK's so-called cheap food. In 1996, the Department of Health calculated that:[12]

- heart disease drugs cost the National Health Service (NHS) £500 million per annum;
- bowel cancers cost £1.1 billion; and
- diseases of the circulatory system cost 12.1 per cent of total health and social services budget.

The British Heart Foundation Research Group at Oxford calculated the costs of coronary heart disease (CHD) for 1993–94. In summary these were:[13]

- 66 million lost working days, equivalent to 11 per cent of all days lost due to sickness;
- £858 million in invalidity benefits;
- £3 billion for lost production in British industry;
- a minimum of £1420 million in direct treatment costs to the NHS; of this the bill for drugs alone was £650 million (there are 23,000 coronary by-pass operations annually); and
- total costs of approximately £10 billion per year; including NHS costs.[14]

The UK's food bill is about £60 billion a year but it should be increased to take account of heart disease costs. And diet is estimated to account for about one third of all heart disease.[15] This not only means taxpayers are spending more money to compensate for failures in the food system but also that human suffering is not included in the costs. Social and monetary costs go hand in hand.

Food poisoning is another externalized cost – estimated at £1 billion per year for the UK.[16] The care cost to the NHS for 100,000 people treated in 1991–94 was £83,139,685.[17] Food poisoning figures, whether reported or proven, are also rising inexorably. One possible major cause for this rise is that food may be more contaminated – due to cross-contamination or to more pathogens in the 'reservoir' or even to the complexity of the food system. The other major cause could be changed consumer behaviour due to new technologies, shopping patterns or lack of knowledge.

The latter interpretation, which has often been promoted by food industry interests understandably wishing to divest themselves of responsibility, has an interesting sting in its tail. If consumer responsibility is generally accepted, what happens when it can be proven that the consumer was *not* responsible? In the US, where the consumer culture of using litigation is furthest advanced, one estimate for the annual cost of food-borne illness from pathogens ranges from $7.7 billion to $8.4 billion.[18] Each case of *salmonellosis*, for instance, is $500–$1350, with a case of botulism working out at $322,000. Unsurprisingly, legal costs account for a heavy proportion of this amount. The point for my present case is that it is the consumer who is paying that cost. Even if it is insurance companies who shell out, their costs are recouped through future premiums. So the food industry is not paying the cost.

ENVIRONMENTAL COSTS

Environmental costs are externalized, too. In the UK, water consumers are paying for the cost of cleaning up pesticide residues that get into drinking water from 'efficient' farming.[19] Ofwat, the water industry regulator, calculates that the capital costs of installing activated carbon to reduce residues to permitted levels is £1 billion.[20] The combined annual capital and running costs of reducing pesticide residues is £100 million.

Distribution, not just production, carries environmental costs. Today in the UK, the same amount of food is essentially travelling further. Basically, the British are eating a constant amount but over the last 15 years the distance this tonnage is transported has gone up by a third, increasing food miles.[21] Not only is the food travelling further, but we are travelling further to get it and using cars more to do so. The distance the British travel for shopping in general rose by 60 per cent between 1975/6 and 1989/91, but the travel by car more than doubled.[22] We are also travelling further to get to the shops. The way in which food is grown and what is grown also have an immense impact on the energy used (see Figure 12.2). In winter, it takes about 4kg of diesel to produce 1kg of greenhouse vegetables but in summer it is less than half of that. It also takes almost as much to produce veal as growing vegetables in winter.

Our production methods also threaten biodiversity. A huge amount of the vegetables grown, for example, are from a narrower and narrower genetic base. More and more of our land is devoted to mono-cropping. The top variety in winter wheat accounts for over one fifth of the land put down to winter wheat in the UK. The top variety sown of oats accounts for nearly half of the oats grown by hectarage.[23] This is a very narrow base and greatly reduces the biodiversity of our food supply.

QUESTIONS OF CONTROL

The continuing concentration of economic power in the food system is changing who controls our food supply. Mergers and acquisitions abound in the food sector throughout the world (see Table 12.1) In Western Europe alone there were 216 mergers and acquisitions in 1993–95. The world total was 533. Already big companies are getting bigger.

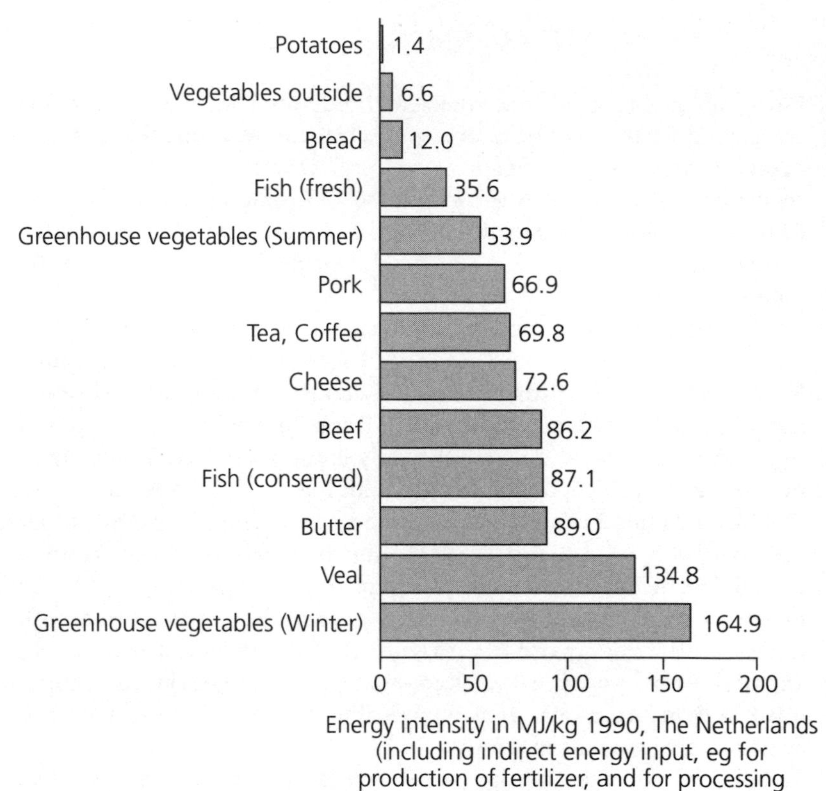

Source: Wuppertal Institute, in Michael Carley and Philippe Spapens, 1998, *Sharing the World*, Earthscan, London

Figure 12.2 ***The Energy Intensity of Food***

In the ice-cream market, for example, Unilever, Nestlé, Haagen-Dazs (a UK company), and Mars (American) dominate. Unilever has 50 per cent of the ice-cream market in the UK, 48 per cent in Germany, 27 per cent in Italy, 37 per cent in the Netherlands, 25 per cent in Spain and 34 per cent in France. In every industry, concentration is the name of the game.

Exactly the same is happening in the US. Table 12.2 shows the concentration ratios of the top four companies in various sectors. Just four companies have 87 per cent of the beef packing market, 60 per cent of pork packing, 45 per cent of broilers and 30 per cent of eggs.

Despite the mythology of choice in the contemporary system, in reality there are fewer companies to choose from. There may be 25,000

Table 12.1 *Number of Mergers and Acquisitions by Main Region, 1993–95*

Region	*1993*	*1994*	*1995*
Western Europe	216	220	260
Central/Eastern Europe	46	29	31
North America	150	138	167
Latin America/Caribbean	37	36	35
Asia/Pacific	71	43	40
Africa/Middle East	13	15	12
World Total	533	481	545
UK total	73	75	12

Source: M Heasman, Thames Valley University/WLTEC

items to pick from as you walk down the aisles of a large supermarket but you are buying from the same few companies. My colleague Dr Michael Heasman of the Centre for Food Policy found that:[24]

- the UK food industry is one of the most concentrated in Europe;
- three companies in 1995 – Unilever, Cadbury Schweppes and Associated British Foods – represented two thirds of total capitalization in UK food manufacturing;
- these companies compete on the world stage and their plant investment decisions involve comparisons between locations able to serve the whole European market;
- of the top 50 European companies, 19 are UK companies;
- UK companies are second only to those of the US in the level of their foreign direct investment in other countries;
- half of the world's top 100 food sector companies are US owned;
- currently, the top 200 groups have combined food and drink sales of £700 billion – broadly half the world's food market; and
- industry privately estimates the global food industry will come to be dominated by up to 200 groups, which will account for around two thirds of sales.

These structural changes in the food system are being accompanied by an astonishing fragmentation of culture. There is a hot debate now, I am glad to report, about whether food culture is subject to deskilling.[25] Because they do not own the land, millions not only do not know how to grow food but also now do not know how to cook it. You might argue that this does not matter. I think potentially it does. I

Table 12.2 *Concentration of US Agricultural Markets*

Food sector	*Concentration ratio (%)*	*Date*	*Top Four Companies*
Beef packers	87	1995	IBP, ConAgra, Cargill, National Beef
Pork packers	60	1997	Smithfield, IBP, ConAgra, Hormel/Cargill
Broilers	45	1996	Tyson, Gold Kist, Perdue Farms, ConAgra
Eggs	30	1992	Cal-Maine, Michael Foods, Rose Acre Farms, Agri-General
Turkeys	35	1992	ConAgra, Rocco, Hormel, Carolina/Wampler-Longacre
Sheep slaughter	73	1996	ConAgra, Superior Packing, High Country, Denver Lamb
Multiple elevator	24	1997	Cargill, ADM, Continental, Bunge
Flour milling	62	1994	ADM, ConAgra, Cargill, Cereal Fd Processors
Soya bean crushing	76	1990	ADM, Cargill, Bunge, Ag Processing

Source: Heffernan, Constance, Gronski and Hendrickson, *Concentration of Agricultural Markets*, University of Missouri, March 1997

do not want to see everyone forced to cook, certainly not women forced into the kitchen just when they have escaped, but I do think it makes sense at least to know the basic skills so that you have the choice. In the UK, cooking came off the curriculum in the mid 1990s. The National Curriculum has no component giving practical food skills. In classrooms we now teach children theoretical exercises about designing food wrapping, rather than how to cook and handle food. Over 80 per cent of food is bought in some way pre-processed, yet we are all encouraged to take more control over our food. The reality is that cooking now occurs mainly in the factory or commercial kitchen.

I am not intrinsically against ready-made foods. The issue is whether, if we do not educate people to be able to create a quick simple meal themselves, we are not creating a new dependency culture. It makes little sense to urge people to look after themselves more, on the one hand, and to deny them the skills with which to do so, on the other. There is already evidence of a schizophrenic culture where there is a skill-rich and a skill-poor class division occurring. A 1993 study found that 93 per cent of UK 7–15 year olds could play computer games, but less than 40 per cent could bake a jacket potato.[26] Even the last government recognized something was wrong there. Unfortunately, the curriculum has not been altered, but some good projects have been set up.

COMPLEX TRENDS, DIVERSE ACTIONS

As passionate observers and actors within the food system, we are faced with a paradox. The food system is complex but we try to make sense of it with simple messages. The food miles argument, for instance, came out of years of tussling with the issue of externalities. We knew about the energy infrastructure of modern food distribution but needed a simple way of communicating and highlighting this issue for consumers. The notion of the food mile was invented to convey these complexities in a way that could make sense to people.

Food, as we know, is a highly ideological topic. There are contrary pressures and trends within the food economy. Table 12.3 summarizes some of these counter trends. I think it is useful to see the food world in this way because history suggests that the future is remarkably plastic. With so many contradictory forces, the direction of world food is uncertain. Globalization is tugging in one way and localization in another; top-down decisions may conflict with bottom-up pressures from people demanding more involvement; intensification continues apace while there is growing pressure to 'extensify' production; and so on. The future is heralded by many questions. Is food coming to the people, or people to the food? Will food come from factories or from the land? Is it monoculture or biodiverse? Is it standardized foods, global food, burger culture food, or are we really going to have difference, cultural difference, biodiverse difference? Do we talk about fragmented cultures or common cultures, and so on? Perhaps the biggest question is how we think of ourselves. Do we think of ourselves as consumers of food, or do we think of ourselves as citizens? Citizenship, central to European culture since the 1789 French Revolution, is perhaps the unifying idea. Its thesis – aspiration – is that we are all participants in, actors in, potential accountees of, a global food system.

Table 12.3 ***Different Futures: Dimensions of Change in the Food System***

globalization	vs	localization
urban/rural divisions	vs	urban-rural partnership
long trade routes (food miles)	vs	short trade routes
import/export model of food security	vs	food from own resources
intensification	vs	extensification
fast speed, pace and scale of change	vs	slow speed, pace and scale of change
non-renewable energy	vs	reusable energy
few market players (concentration)	vs	multiple players per sector
costs externalized	vs	costs internalized
open farming systems	vs	closed systems
rural depopulation	vs	vibrant rural population
monoculture	vs	biodiversity
agrochemicals	vs	organic/sustainable farming
biotechnology	vs	indigenous knowledge
processed (stored) food	vs	fresh (perishable) food
food from factories	vs	food from the land
hypermarkets	vs	markets
de-skilling	vs	skilling
standardization	vs	'difference'
superficial variety on shelves	vs	real variety on field and plate
people to food	vs	food to people
fragmented (diverse) culture	vs	common food culture
created wants (advertising)	vs	real wants (learning thro' culture)
'burgerization'	vs	local foods specialities
microwave re-heated food	vs	cooked food
global decisions	vs	local decisions
top-down controls	vs	bottom-up controls
dependency culture	vs	self-reliance
health inequalities widening	vs	health inequalities narrowing
social polarization and exclusion	vs	social inclusion
consumers	vs	citizens

THE WAY AHEAD

I have long argued that consumers and campaigners do not give enough time to understanding the drivers of the food system. In particular, we are poorly researched on the new baronial class that controls the food economy. Do we really know enough about the companies that straddle the world? It is hard to study them, but if we do not, we are colluding in our own deficiencies. I have learned that from an area of particular importance to me – food poverty in Britain.[27] I spend a considerable amount of my time looking at the food retailers because their actions impinge on the 'freedom' of people on low incomes particularly directly. I have learned of the power of retailers not just through the contracting process but the siting of stores.[28] Such understanding can help those concerned about animal welfare to appreciate potential levers for change in their area too.

That said, our understanding has improved in leaps and bounds these last two decades. We are pretty clear that what is needed to improve the lot of animals, humans, land and all life requires a diversity of action – but a range of actions that are co-ordinated. Diversity, with a coordinated message, is much more appropriate for bringing about real change.

A TIME FOR ACTION

Now is a good time for action. We have been marking the 50th anniversary of the Universal Declaration of Human Rights, the 50th anniversary of the NHS in the UK and the 150th anniversary of the first Public Health Act recently. These things did not just happen. They were fought for by a wide range of people. Today, many people are beginning to question and ask long-term questions. What's life for? How much money do I really need? How much work do I really want to do, if it means I do not see my children? Questions about animal welfare are part of this wider cultural questioning.

Looking ahead, there are many interesting ideas for us to pursue. On globalization and investment generally there is the Tobin Tax – a suggestion by Nobel prize-winner James Tobin, who argues that there should be 0.5 per cent tax on transactions on the global financial markets. The transnationals do not want it, the banks do not want it. But people who want to limit speculative investment and link investment to its effects on people's lives do. With currency markets in some difficulty from the Far East to Russia, perhaps this is an idea whose time has come. What is its relevance to animal welfare? In a world where governments and neo-liberal thinking say there is not enough money to be able to afford improvements in welfare or environmental protection, Tobin reminds us that the world is in fact awash with capital; it is just that most of it never 'lands'.

In food policy, I see competition policy as a crucial area in the future. There has been an extraordinary concentration of control and market share. In food retailing, four companies now sell over 60 per cent of all food in the UK – an *astonishing* state of affairs – but competition policy says a company is not a potential monopoly until it has 25 per cent of the market. The big multiple retailers can rightly argue that they only have, say, 10–15 per cent, but if we look at the south of the UK, we find that that Sainsbury's and Tesco probably have well over 25 per cent. In short, the policy question is: what is the geography of a market? Is it five miles from where a consumer lives? 20? 100? The UK? Europe? The world? By defining markets, big companies are always in the position to de-emphasize their power and influence. Who defines this market and competition policies in it? Some serious economic, economic policy and competition policy work needs to be done on this. This is a key issue through which we can get a handle on this otherwise seemingly enormous power that can make us feel *powerless.* The rhetoric of our society says we live in a market economy with free choice. The reality is less simple.

Notes

1 C P Timmer, W P Falcon and S R Scott (1983) *Food Policy Analysis*, World Bank, published by Johns Hopkins University Press, Baltimore, p 9, quoted in G Tansey and T Worsley (1995) *The Food System – a Guide*, Earthscan, London
2 OECD (1981) *Food Policy*, OECD, Paris
3 T Lang (1997) *Food for the 21st century. Discussion Paper 4*, Centre for Food Policy, Thames Valley University, London
4 See R Wilkinson (1996) *Unhealthy Societies*, Routledge, London
5 For details about food poverty in a rich country like Britain, read the government report from the Food and Low Income Working Party: LIPT (1996), *Low income, food, nutrition and health: strategies for improvement*. A report by the Low Income Project Team for the Nutrition Task Force, Department of Health, London, or S Leather (1996) *The Making of Modern Malnutrition*, Caroline Walker Trust, London
6 Nutrition programme, WHO Regional Office for Europe, Copenhagen
7 EC figs: quoted in M Whitehead and P Nordgren, eds (1996) *The Health Impact of CAP*, National Institute of Health, Stockholm
8 A Paxton (1994) *The Food Miles report*, Sustainable Agriculture, Food and Environment (SAFE) Alliance, London
9 D MacLaren, S Bullock, N Yousuf (1998) *Tomorrow's World: Britain's share in a sustainable Future*, Earthscan, London, 149
10 G Conway (1998) *The Doubly Green Revolution*, Penguin, Harmondsworth
11 J Kuusipalo, M Mikkola, S Moisio, P Puska (1988) 'Two years of the East Finland Berry and Vegetable Project: an offshoot of the North Karelia Project' *Health Promotion International*, vol 3, 3, pp 313–317

12 NHS Executive (1996) *Burdens of Disease*, Department of Health, London
13 1993–94 figures. British Heart Foundation (1996) *Coronary Heart Disease statistics*, BHF, London
14 BHF (1998) *CHD statistics – Economic Costs*, British Heart Foundation Research Group, Oxford
15 C Williams (1998) *Food, Nutrition and Cardiovascular Disease Prevention in the European Union*, European Heart Network, Brussels
16 Jenny Roberts (1995) 'The socio-economic costs of foodborne infection', unpublished paper to Oxford Brookes University conference 'Foodborne disease: consequences and prevention' St Catherine's College Oxford, April
17 CDSC (1995) 'The cost of in-patient care for acute infectious intestinal disease in England, 1991–1994', *Communicable Report*, vol 6, Review no 5
18 *J of Environmental Health*, Sept/Oct 1992
19 G Conway, J Pretty (1991) *Unwelcome Harvest: Agriculture and Pollution*, Earthscan, London
20 Ofwat (1997) 1996–7 Report on the financial performance and capital investment of the water companies in England and Wales, Ofwat, London
21 A Paxton (1994) *The Food Miles Report*, SAFE Alliance, London
22 H Raven, T Lang (1995) *Off our Trolleys?*, Institute for Public Policy Research, 8, London
23 R Jenkins (1992) *Bringing Rio Home: biodiversity in our food and farming*, SAFE Alliance, London
24 M Heasman (1997) *Getting a quart out of a pint pot: ownership and restructuring in UK food production*. Paper to GMB conference, Unpublished, Centre for Food Policy/West London TEC, London
25 Caraher, M, Dixon, P, Lang, T, Carr-Hill R (1998) 'Barriers to accessing healthy foods: Differentials by gender, social class, income and mode of transport' *Health Education Journal*, September; T Lang, M Caraher (1988) 'Food poverty and shopping deserts: what are the implications for health promotion policy and practice?' *Health Education Journal*, September
26 NFA (1993) *Get Cooking!*, National Food Alliance and Department of Health, London
27 T Lang (1997) 'Dividing up the cake: food as social exclusion' in A Walker and C Walker, eds, *Britain Divided: the growth of social exclusion in the 1980s and 1990s*, Child Poverty Action Group, London
28 H Raven, T Lang (1995) *Off our Trolleys? Food retailing and the hypermarket economy*, Institute for Public Policy Research, London

Part V

Farming Threats and Opportunities

Today enormous forces are trying to reshape our food system. Technological changes are underway which could completely transform the crops and animals produced over the next 20 years while alternative methods are neglected (Chapters 13–15). Even those committed to many alternatives frequently fail to see the possibility of an agriculture not built around killing animals, as is argued in Chapter 16.

13 Genetic Engineering and the Threat to Farm Animals

Joyce D'Silva

In 1985, the world looked on horrified as scientists at the US Department of Agriculture's (USDA) Beltsville research centre called in the media to see the first ever GE farm animals – creatures that became known as the Beltsville pigs. These pigs had been genetically engineered with growth hormone genes taken either from humans or from cattle, in an attempt to produce leaner, faster-growing animals.

In their euphoria at creating the first ever transgenic livestock, the genetic engineers had not anticipated public reaction to pictures showing these pigs, many of whom had damaged vision, deformed skulls and were unable to walk properly. The pigs also displayed increased susceptibility to pneumonia and gastric ulcers.[1]

It was a scientific triumph, and an animal welfare and PR disaster. Beltsville went quiet. But do not be deceived. The work goes on. The only difference is that the cameras are no longer called in to film the transgenic animals.

In 1992, a Beltsville paper records how a gene usually found in a chicken – the cSKI gene – was micro-injected into pig ova. The scientists had already put this gene into mice and observed a huge increase in muscle and a 'near absence of body fat'. Now they were aiming for the same effect in pigs. Five transgenic pigs did develop varying degrees of increased muscular growth between three and seven months of age. But five other transgenic piglets developed floppy, weak muscles in their legs between birth and three months of age. The scientists concluded that this gene 'may profoundly affect muscle development'.[2] This time the media were not called in. There was no published video of piglets unable to stand up and walk properly.

Sometimes, it seems, genetic engineers are slow to learn from their mistakes. Not long after this disaster, genetic engineers at Colorado State University transferred a similar chicken SKI gene into calf embryos. One transgenic bull calf was born. For eight weeks he grew normally, then he began to show increased muscle on his loins and rear quarters. At ten weeks he started to get weak and became unable to stand without assistance. Tests showed muscle degeneration. The scientists tell us:

> *Throughout this period, the bull remained alert, manifested no signs of pain and retained a normal appetite. At 15 weeks of age, it became obvious that the muscle weakness was not a transient phenomenon and humane considerations led us to euthanase this animal (and for control purposes his twin).*[3]

An Australian project to improve wool growth put (bacterial) cysteine (biosynthesis) genes into lamb embryos. The transgene was found in only two out of 46 lambs born live and in ten out of 20 that died. The scientists concluded that their experiment 'may have a deleterious effect on the physiology of developing lambs'.[4]

In 1997, roughly the same group of scientists who produced the Beltsville pigs published a paper on the transfer of a sheep growth hormone gene into pigs. Fifteen piglets were born transgenic – three were either still-born or died. This particular gene was supposed to be 'turned on', as it were, by zinc in the diet. Indeed, there was a huge increase in the levels of growth hormone in the plasma of the transgenic pigs after zinc was added to their feed. In one female pig, No 7203, there was a near 20-fold increase in growth hormone levels. The day after the zinc supplement was stopped, she died. The scientists wrote, 'Death resulted from acute gastric hemorrhage due to ulceration (of the pars esophagea).' They surely were not surprised. Their own previous experiments had shown a high incidence of similar ulceration in the original Beltsville pigs with growth hormone genes and also in pigs injected with porcine growth hormone, PST.

The Beltsville boys concluded, 'The development of a controllable transgene expression system that functions in farm animals is still greatly needed.'[5] Is it? At this cost in terms of animal suffering?

FISH AND FOWL

Transgenic poultry and fish are also being developed. Work with fish is comparatively advanced. As the world's oceans are already being fished to capacity, many see aquaculture – fish farming – as the way to produce animal protein on a vast scale. As one American scientific paper declares, 'The application of biotechnology to aquaculture could provide a competitive edge with the potential to greatly reduce the US trade deficit.'[6]

A 1994 review paper from scientists in India refers to many genetic engineering experiments with fish including several where the human growth hormone gene has been transferred to Atlantic salmon, trout, tilapia, goldfish, loach and catfish.[7] The race is on not only to produce fish which grow bigger but also fish which have antifreeze genes like the winter flounder which can live in waters of –2°C. Scientists hope that transgenic farmed salmon will be able to survive winter in salt water aquaculture tanks in cold regions.[8]

But even with fish, things can go wrong. In an experiment to put an extra growth hormone gene from sockeye salmon into coho salmon, on average the transgenic salmon were 11 times heavier than the non-transgenic control salmon. One salmon was 37 times larger. In a further experiment, this group of genetic engineers put a growth hormone gene into coho salmon and found one group of salmon hatched from transgenic parents developed deformed heads and a green discoloration. After one year the overgrowth of cartilage 'became progressively more severe and resulted in reduced viability'.[9]

The failure rate of genetic engineering of farm animals is enormous – both because of the hit-and-miss nature of the technology itself and because of its effects on animal well-being.

CLONING

Now a new biotechnology is at hand – cloning. This may even provide a technological fix for the failures of genetic engineering. Even if you only produce one viable transgenic pig from the kind of experiments described above, by cloning it you can have lots of transgenic pigs. Already the team at the Roslin Institute at Edinburgh has produced Polly – the first farm animal to be both cloned *and* genetically engineered.[10]

Cloning is bad news for animal welfare. Polly and Dolly were preceded by Megan and Morag – the two first cloned lambs produced at Roslin. Although public attention was focused on the success of the experiments, three other cloned lambs were born in the same batch – two died shortly after birth, the other at ten days. CIWF has long suspected that many scientific papers fail to tell the whole story. They are explicit on scientific procedure, but neglect to tell all about the full scale of failures. In the case of Megan and Morag, the published paper in *Nature* magazine failed to mention that the three lambs who died all had malformed internal organs. It also failed to mention the abnormally large birth weights of the lambs.[11]

Yet abnormal birth weights have been a virtually consistent feature in cloning experiments with calves. In 1996, Colorado State University research reported production of 40 cloned calves, many of which showed abnormal birth weights. The heaviest calf (67.3kg) could not stand without external support. Six other calves also had limb contraction problems. Thirty-four of the 40 transgenic calves required medical therapy for conditions ranging from weakness and hypothermia to hypoglycemia. Eight died. The scientists write, 'These cloned calves typically did not exhibit normal behaviour patterns and would be characterised as slow and/or weak.'[12]

But the experiments go on. In January 1998, a New Zealand team announced the birth of seven cloned lambs, but only two survived. The others lived for between ten minutes and seven days 'with lung dysfunction being a major mortality factor'.[13] At Roslin, having cloned Megan and Morag from sheep embryo cells, the team declared a world first in February 1997 with the announcement of the 'creation' of Dolly – the first mammal cloned from a cell taken from an adult, in her case, an udder cell from another sheep. Seven other clones were born at the same time, of which one died soon after birth.[14]

A year later Roslin announced the birth of Polly, the first sheep to be both cloned and genetically engineered. Six cloned lambs were born alive, one died. One foetus that died *in utero* was malformed. But to the scientific world this was a day of optimism. 'The realistic prospect of targeted genetic manipulation in a livestock species should open a vast range of new applications and research possibilities,' the team wrote.[15]

The Roslin Institute is closely connected to PPL Therapeutics – a company set up to commercialize the Roslin work. PPL Therapeutics' press release speaks of the 'commercial potential' of the Polly technology.[16] Polly carries the human gene for the blood-clotting agent, Factor IX. The hope is that she will express this factor in her milk. The *British Medical Journal* says the aim is to induce premature lactation in Polly, so they can test her milk to see if it contains Factor IX.[17]

It is not the first time that a farm animal has had lactation induced. In February 1997, PPL Therapeutics' American arm (PPL Therapeutics Inc.) introduced us to Rosie, a so-called transgenic 'cow' carrying a gene for a human milk protein. Further information revealed that Rosie was only a 6-months-old heifer, so CIWF was surprised to hear she was producing 2.4 grammes of this protein in every litre of milk. Like other mammals, heifers do not produce milk until they have given birth – usually at around two years old. They do not even reach puberty until 15 months of age. We phoned PPL Therapeutics Inc. to be told by their General Manager that Rosie had indeed been induced to produce milk at only six months of age, by injecting her with hormones.[18] The scientific paper bears this out. It looks like this could be Polly's fate also.

Cloning, linked with genetic engineering, is obviously taking off. George and Charlie are two calves, the first to be, like Polly, both genetically engineered and cloned. They were announced to the world in February 1998 by US researchers.[19] We await the published paper.

One of the world's leading experts in cloning, George E Seidel Jr, reckons that abnormalities occur in 1 per cent of calves produced by natural mating, 10–15 per cent of calves from embryos cultured in vitro and close to 50 per cent with calves cloned by nuclear transplantation.[20]

Although the experiments to induce increased growth rates and muscle development continue, much of the focus of farm animal genetic engineering is on the production of specific proteins in milk. If the animal produces the desired protein then it could be purified from the milk.

Experiments have been going on since the mid 80s. As usual, there have been animal casualties along the way. A 1991 Beltsville experiment in transferring a mouse whey gene into pigs did result in pigs who expressed the protein – the mouse whey acidic protein – in their milk. But five out of eight transgenic sows stopped producing milk by the tenth day of lactation. As usual, the story was sold as a success, with the researchers concluding optimistically, 'As shown here, it is possible to convert transgenic pigs into bioreactors producing and secreting into milk large amounts of non-porcine proteins.'[21]

The proponents of this pharmaceutical 'pharming' always tell us that as the transgene is only produced in the udder, it cannot affect the health of the transgenic animal. It would seem they could be wrong.

Ian Wilmut, the leader of the Roslin team who created Dolly, has admitted that sometimes the expression of the transferred protein gene occurs outside the udder. As he says, 'Non-specific expression in this way could be very damaging... Production in the plasma of proteins such as clotting factors or erythropoietin even at modest concentrations could be harmful to the animal.'[22] Another noted French researcher in this field, L M Houdebine, also writes that 'the health of the animals may be severely altered by the recombinant proteins transferred from milk to blood'.[23]

But if not milk, what about blood? Several researchers have produced transgenic pigs in a bid to produce, for example, human haemoglobin in their blood. A scientist working for DNX Biotherapeutics (now known as Nextran), declared, 'since large amounts of blood can be harvested during the lifespan of a pig, as much as 1kg. of pure human haemoglobin can be extracted from just one such transgenic pig'.[24] How would the blood be harvested? Beltsville's Robert Wall suggests that blood 'could be harvested by catheterization or at slaughter'.[25]

If blood presents problems, then what about urine? In January 1998, *Nature Biotechnology* magazine carried a paper, from Beltsville again, about the production of human growth hormone in the urine of transgenic mice. The first five female transgenic mice were all sterile, but they were able to breed further transgenic animals from the males. The human gene product was found in the urine – although it was also expressed in the brain, kidney and testes. These researchers envisage herds of transgenic cattle producing products such as human blood-clotting Factor VIII in their urine.[26] The animals, however, would have to be immobilized all their lives attached to catheters to extract their precious urine. As the Beltsville researchers acknowledge, 'A major advantage of bladder production of pharmaceuticals is the ability to harvest the product soon after birth and throughout the life of the animal (without regard to its sex or reproductive status).'[27]

The technology

To produce transgenic or cloned animals, scientists use a range of reproductive technologies. To obtain ova for micro-injection or for cloning, the female animals are given repeated hormone injections to induce super-ovulation – the production of many more ova than usual. The animals are artificially inseminated if embryos are required. Then a laparotomy operation is performed – a cut through the side of the animal to enable the oviduct to be pulled out of the abdominal cavity to flush out the ova. The operation tends to result in adhesions which, as one researcher puts it, 'limit the use of donor animals'.[28] In the laboratory, the tiny zygotes (fertilized ova) are then micro-injected with hundreds of copies of the desired gene.

A new technique, ovum pick-up, is now being used in cattle. Using ultrasound, a needle is guided through the cervix and punctures the follicles to collect the ova. This saves the surgical operation, but means the animals can be subjected to this distressing procedure on a regular basis. One team recommended that it 'can be performed twice weekly yielding 5–7 zygotes per session (40–56 zygotes per donor in a 2-month period)'.[29]

For cloning, the nucleus of the egg is removed and the new genetic material, usually taken from the foetus of another animal, is fused into the enucleated ova via an electric pulse. In both genetic engineering and cloning, the tiny embryo is then transferred via surgery to the oviduct of another animal, whose oestrus cycle has been synchronized with hormone injections. Frequently a temporary surrogate mother is used. This could be a rabbit or a sheep. She is killed after about a week and the embryo is removed and checked for viability before being implanted – again usually by surgery – into the final surrogate mother, who carries the animal to term. Birth is sometimes natural. As clones often show excessive growth *in utero*, a caesarean is often required. Sometimes synchronized caesareans are performed anyway to avoid any problems which may arise at birth. So, for example, to obtain Dolly, three surgical operations and one killing took place – and she was not born by caesarean.[30]

One reason why genetic engineering experiments are nearly always performed first on a multitude of mice is to cut the costs of rearing large mammals. Cattle, in particular, take two years to reproduce, which is a long time for the genetic engineers to wait to see if a transgene has been passed on to a subsequent generation.

Now scientists in Australia have stimulated follicle development and produced mature egg cells in calves as young as three weeks old by giving them a series of hormone injections. The calves were then subjected to laparoscopic surgery to observe and/or remove the ova for in vitro fertilization. Many of the calves were subjected to the same hormone treatment and repeated operations at six and nine weeks of age.[31] Similar experiments have resulted in ova collection from hormone-treated lambs aged only 8–9 weeks old.[32] Oocytes (ova before they mature and begin to divide) could be removed from calf foetuses whilst they are still in the uterus.[33] This technique would entail obtaining foetal oocytes as early as 90–180 days into gestation and using in vitro fertilization in the laboratory.[34] The oocytes could also have genes added at this stage. A whole new science is envisaged which, one researcher claims, would lead to 'the rapid and efficient intraspecies transfer of desirable genes (between genetic backgrounds)'.[35]

Genetic engineering and cloning are still in the developmental stage – and are using an increasing number of farm animals. In the UK, between 1990–97, the number of GE animals used in scientific procedures increased by 730 per cent. Many of these animals are mice, but in 1997, 1,472 sheep were used in genetic engineering experiments, a one third increase over the previous year.[36]

Some of these animals are being used to replace mice as models of human diseases. Dr Griffin from the Roslin Institute points out that the lung physiology of sheep is closer to humans than mice lung physiology and therefore transgenic sheep are likely to be better models for work on cystic fibrosis. But, as he admits, 'The deliberate creation of a genetic disease in a large animal raises public concerns and the ethical justification of creation of such a model would need to be well argued.'[37]

The huge number of transgenic mice being created has led to the setting up of the National Repository for Transgenic and Targeted Mutant Mice in the US. In Europe, the European Mouse Mutant Archive is also being formed. John Sharp of the Jackson Laboratory says, 'It is clear that the number of transgenic animals is also growing exponentially.'[38]

Why is there such a huge increase? For the scientists the motivation could be what Dr Harry Griffin, from the Roslin team, calls the excitement of 'being part of the show'.[39]

APPLYING THE TECHNOLOGIES

These technologies can be applied in various ways. The Belgian researcher, Michel Georges, tells of the importance of experiments in which newly discovered genes are 'knocked out' – the resulting animal will then show what is missing, as it were. Whilst admitting most of these animals are debilitated, he predicts a dramatic increase in such experiments as they may 'point towards genes of potential interest to animal breeders and deserving further manipulation in livestock'.[40]

Dr Griffin from Roslin speculates that farmers could use embryos cloned from the most productive cows to lift their own herd's performance. He admits there are major technical barriers to be overcome first – and reckons this will take 10–20 years. In other words, 10–20 further years of hit-and-miss technology applied to living creatures. Ultimately, he reckons cloning could, in principle at least, produce 'unlimited numbers of genetically identical animals' as well as 'a route to gene targeting in livestock'.[41]

Colin Stewart, writing in *Nature* after the news of Dolly, says, 'Maybe in the future the collective noun for sheep will no longer be a flock – but a clone.'[42]

Many are still speculating about the development of fast-growing, high muscle, low-fat farm animals. Through selective breeding we already have so-called 'double-muscled' breeds of cattle like the Belgian Blue. These animals develop huge meaty hind quarters which become profitable meat. It has recently been discovered that the cause of this increased muscle is a mutation in the myostatin gene in these animals. Now American researchers at Johns Hopkins University have speculated, 'It may be possible to produce double-muscled sheep, pigs, fowls, turkeys and fish by introducing mutations in the myostatin gene.'[43] This is despite the fact that cows pregnant with Belgian Blue calves often require caesareans[44] and their calves have reduced viability.

Genetic engineers also aim to produce GE animals resistant to disease. Of course, the diseases are likely to be those diseases endemic in factory farming systems. But as one leading authority says:

> *selection for increased immunity to one disease will likely result in decreased immunity to other diseases... if one selects for too much immunological activity, there will likely be an increase in autoimmune diseases in addition to interference with normal functions such as reproduction.*[45]

The major growth area in farm animal genetic engineering at the moment is the production of transgenic animals which produce a specific product in their milk. Predictions include a world market for pharmaceutical products produced in transgenic animals of over $3 billion a year.[46] Already Nexia Biotechnologicals (Canada) has opened a transgenic animal facility with the aim of housing 1,000 GE goats in the near future. The goats will have genes enabling them to produce human proteins in their milk.[47]

Yet, 'The problem of the possible presence of agents pathogenic for humans in proteins extracted from milk is not completely solved,' writes the leading French scientist L M Houdebine.[48] This note of great caution is echoed by Steven Bauer from the Food and Drug Agency Centre for Biologic Evaluation, who says, 'The potential use of a wide variety of new animal hosts for production of therapeutics presents the possibility of exposure of humans to new pathogens.'[49]

Eating GE animals

Will you be eating transgenic animals themselves? In 1992, in the US, 66 farm animals, who had been micro-injected with foreign genes as embryos but who appeared not to have incorporated the transgene, were slaughtered for human consumption. In the US, the Agricultural Biotechnology Research Advisory Committee has produced guidelines on the food safety of products from transgenic animals themselves. The Committee decided that 'if there are no effects of food safety concern, the transgenic animals and their progeny may enter the human food chain'.[50] So American consumers may already be eating transgenic farm animals.

In the UK, the Advisory Committee on Novel Foods and Processes has also said it sees no problems in allowing the failures of transgenic experiments into the food chain. The Food Advisory Committee, however, has said their initial view is negative, but that if the meat is given the go-ahead it should be clearly labelled.[51]

Genetically engineered creatures released onto farms and into fish farms could pose threats not just to consumers' health. Giant transgenic or ice-compatible salmon may compete with fish already existing in the fragile oceanic ecosystem. Fields, or more likely sheds, of cloned transgenic farm animals may lead to a huge decrease in genetic diversity. Around one third of the earth's 4,000 breeds of domestic animal species are at risk of extinction according to the FAO.[52] Also, with a herd of identical cloned animals, all could be identically vulnerable to the same disease pathogen.

It also seems unlikely that genetic engineering will be good news for farmers in developing countries. Their own hardy native breeds of farm animals will be replaced with the transgenic clones. There will be patent fees to pay. The richer farmers could get richer, the poor more marginalized than ever.

Greatest risks

The greatest risk to farm animal welfare is not that the genetic engineers will fail, but that they will succeed. Once the success rate starts outstripping the failure rate, the whole business will explode. Then experiments to find the best gene for fastest growth, or the highest milk yield, will really take off, and thousands more transgenic animals will be 'created' for an uncertain future.

There is another, more insidious, threat to these animals – not only will their physiology be altered radically, but so will their brains. There is already serious talk of creating animals whose consciousness is altered so that they no longer resist being caged and crated. Then we can have vegetable-like hens 'happily' existing in tiny cages, and pigs who are infinitely 'content' on concrete behind bars, and probably cows that do not fret when their calves are taken away from them.

Already in *Nature* magazine, we see the genetic engineers taking the first steps on this road with a series of hideous experiments in which mice were genetically engineered to be less sensitive to pain. Of course, to test the success of their experiment, the scientists forced the transgenic mice and ordinary 'control' mice to stand on hotplates and also injected and painted them with an extract from red peppers known to cause a sensation of burning pain – amongst other things.[53]

This represents the total destruction of the essence, of the 'telos', of the animal. It reduces individual sentient creatures to mere functional gene products. As Caird Rexroad puts it, from his genetic engineer's viewpoint 'an animal can be viewed as a set of genetic potentials'.[54] But logically, therefore, that is all *we* are too – sets of genetic potentials. Is that how you and I *feel* – or do we feel, are we not certain, that we are more than that? And if us, surely also them?

Once more we see humanity entrenched in anthropocentrism – seeing everything from our own viewpoint. Animals as mere means to an end.

WHOSE ENDS COUNT?

I am no Luddite. I am not against biotechnology *per se*. Even in animal farming it may have a role in providing better ways to diagnose disease. I do not have an intrinsic objection to its use on milk and meat, once the product has been divorced from the living animal. It is living creatures that suffer, not lumps of flesh or mammary products.

But I do believe that each animal, whether it be just one of the 700 million broiler chickens reared each year in the UK, or the 100 millionth pig in China – each of these creatures is a sentient being, each has a capacity to suffer, each is as unique as you or I.

Already, we exert power and control over their lives and deaths. We condemn them to lives of misery and deaths often of agony. We are abusing our power.

Now science offers us a whole new dimension of power and control – not just over living conditions and slaughterhouse practice –

but over the very physiological make-up, even the mental capacity, of the animals themselves. We can stop them being who they are. We can mutilate their bodies before they are born, we can reduce highly intelligent sentient beings to inert lumps of muscle and blood. We have this appalling power at our fingertips. And I say we must reject it.

Fifty years ago the world began to realize that nuclear weapons gave us a power too great to use. The world has held back from using these weapons ever since.

I call on:

- the scientific community to stand back from exercising their new genetic engineering technology on animals;
- governments to stop funding it and to regulate it out of existence; and
- all of us, as caring human beings, to declare our opposition to this technology and to work hard to halt its development.

It is hard to stop the race once it has begun. But we have another tactic. We can promote the concept that animals are individual sentient creatures, capable of suffering and therefore worthy of our concern and care. That is not a lot to ask. But if that mind-set could prevail, then the cold calculations of the biotechnology companies and the myopic science establishment would find that their work could never ever be justified.

Notes

1 Bolt, D, Pursel, V, Rexroad, C, Jr and Wall, R A, of USDA (1988) 'Improved animal production through genetic engineering: transgenic animals' in Proceedings of forum, Veterinary Perspectives on Genetically Engineered Animals, sponsored/published by the American Veterinary Medical Association

2 Pursel, V G, Sutrave, P, Wall, R J, Kelly, A M and Hughes, S H 'Transfer of c-Ski gene into swine to enhance muscle development' *Theriogenology* 37, p 278, January 1992

3 Bowen, R A, Reed, M L, Schnieke, A, Seidel, G E Jr, Stacey, A, Thomas, W K and Kajikawa, O (1994) 'Transgenic Cattle Resulting from Biopsied Embryos: Expression of c-ski in a Transgenic Calf' *Biology of Reproduction* 50, pp 664–668

4 Bawden, C S, Sivaprasad, A V, Verma, P J, Walker, S K and Rogers, G E (1995) 'Expression of bacterial cysteine biosynthesis genes in transgenic mice and sheep: toward a new *in vivo* amino acid biosynthesis pathway and improved wool growth' *Transgenic Research* 4, pp 87–104

5 Pursel, V G, Wall, R J, Solomon, M B, Bolt, D J, Murray, J D and Ward, K A (1997) 'Transfer of an Ovine Metallothionein-Ovine Growth Hormone Fusion Gene into Swine' *J Anim Sci* 1997 75, pp 2208–2214
6 Powers, D A, Chen, T T and Dunham, R A (1992) 'Transgenic Fish' in *Transgenesis*, Ed J A H Murray, John Wiley & Sons Ltd
7 Pandian, T J and Marian, L A (1994) 'Problems and prospects of transgenic fish production' *Current Science*, Vol 66, No 9, 10 May
8 See ref 6
9 Devlin, R H, Yesaki T Y, Blagi, C A, Donaldson, E M, Swanson, P and Cha, W-K (1994) 'Extraordinary salmon growth' *Nature* Vol 371, 15 September, pp 209–210
10 Schnieke, A E, Kind, A J, Ritchie, W A, Mycock, K, Scott, A R, Ritchie, M, Wilmut, I, Colman, A and Campbell, K H S (1997) 'Human Factor IX Transgenic Sheep Produced by Transfer of Nuclei from Transfected Fetal Fibroblasts' *Science*, Vol 278, 19 December, pp 2130–2133
11 Campbell, K H S, McWhir, J, Ritchie, W A and Wilmut, I, Roslin Institute, Edinburgh (1996) 'Sheep cloned by nuclear transfer from a cultured cell line' *Nature*, Vol 380, 7 March
12 Garry, F B, Adams, R, McCann, J P and Odde, K G (1996) 'Postnatal characteristics of calves produced by nuclear transfer cloning' *Theriogenology* 45, pp 141–152, Elsevier Science Inc
13 Misica, P M, Peterson, A J, Day, A M and Wells, D N, AgResearch Ruakura, Hamilton, New Zealand (1998) 'Co-Transfer of trophoblastic vesicles with cloned sheep embryos is unable to improve survival to term' *Theriogenology*, Vol 49, No 1
14 Wilmut, I, Schnieke, A E, McWhir, J, Kind, A J and Campbell, K H S, Roslin Institute (1997) 'Viable offspring derived from fetal and adult mammalian cells' *Nature*, Vol 385, 27 February
15 See ref 10
16 Dated 24 July 1997
17 *British Medical Journal*, p 169, 17 January 1998
18 Eyestone, W H, Gowallis, M, Monohan, J, Sink, T, Ball, S F and Cooper, J D (1998) 'Production of Transgenic Cattle Expressing Human a-Lactalbumin in Milk' *Theriogenology*, Vol 49, No 1, January
19 AP Health Science 'Praise and Condemnation on Cow Cloning' 20 February 1998
20 Seidel, George E Jr (1998) 'Biotechnology in animal agriculture' *Animal Biotechnology & Ethics*, pp 50–68, Chapman & Hall
21 Shamay, A, Solinas, S, Pursel, V G, McKnight, R A, Alexander, L, Beattie, C, Hennighausen, L and Wall, R J (1991) 'Production of the mouse whey acidic protein in transgenic pigs during lactation' *Journal of Animal Science*, Vol 69, pp 4552–4562
22 Wilmut, I and Whitelaw, C B A (1994) 'Strategies for Production of Pharmaceutical Proteins in Milk' *Reprod Fertil Dev* 6, pp 625–630
23 Houdebine, L M (1995) 'The production of pharmaceutical proteins from the milk of transgenic animals' *Reprod Nutr Dev* 35, pp 609–617, Elsevier/NRA

24 Sharma, A, Martin, M J, Okabe, J F, Truglio, R A, Dhanjal, N K, Logan, J S and Kumar, R (1994) 'An Isologous Porcine Promoter Permits High Level Expression of Human Hemoglobin in Transgenic Swine' *Bio/Technology*, Vol 12, pp 55–59
25 Wall, R J (1995) 'Modification of Milk Composition in Transgenic Animals' Proceedings of Symposium, Beltsville Agricultural Research Center, Beltsville, Maryland, US
26 Kerr, D E, Liang, F, Bondioli, K R, Zhao, H, Kreibich, G, Wall, R J and Sun, T-T (1998) 'The bladder as a bioreactor: Urothelium production and secretion of growth hormone into urine' *Nature Biotechnology*, Vol 16
27 See ref 26
28 Kuhholzer et al (1998) 'Laparoscopic recovery of pronuclear-stage goat embryos' *Veterinary Record* 142, pp 40–42
29 de Loos, F A M and Pieper, F R (1997) 'In Vitro Generation of Bovine Embryos' *Transgenic Animals – Generation and Use*, pp 51–54, Harwood Academic Publishers
30 See ref 14
31 Armstrong, D T, Irvine, B J, Earl, C R, McLean, D and Seamark, R F (1994) 'Gonadotropin stimulation regimens for follicular aspiration and in vitro embryo production from calf oocytes' *Theriogenology* 42, pp 1227–1236
32 Earl, C R, Irvine, B J, Kelly, J M, Rowe, J P and Armstrong, D T (1995) 'Ovarian stimulation protocols for oocyte collection and in vitro embryo production from 8 to 9 week old lambs' *Theriogenology*, 43, p 203
33 Gordon, Ian (1996) 'Controlled Reproduction in Cattle and Buffaloes' Vol 1, p 307, CAB International
34 Georges, M (1991) 'Perspectives for Marker-assisted Selection and Velogenetics in Animal Breeding' in *Animal Applications of Research in Mammalian Development*, Eds Pederson, R A, McLaren, A and First, N L, Cold Spring Harbor Laboratory Press, pp 285–325
35 See ref 34
36 Based on Home Office Statistics of Scientific Procedures on Living Animals, Great Britain 1997, HMSO, 1998
37 Griffin, Dr H 'Cloning and Genetic Modification: Applications in medicine' Roslin Institute, *On Line*, 11 December 1997
38 Sharp, J J and Mobraaten, L E (1997) 'To Save or Not to Save: The Role of Repositories in a Period of Rapidly Expanding Development of Genetically Engineered Strains of Mice' in *Transgenic Animals – Generation and Use*, pp 525 and 530, Harwood Academic Publishers
39 Griffin, Dr H (1997) 'Dollymania' Roslin Institute, *On Line*, April
40 Georges, M (1997) 'Recent Progress in Mammalian Genomics and its Implication for the Selection of Candidate Transgenes in Livestock Species' in *Transgenic Animals – Generation and Use*, p 520, Harwood Academic Publishers
41 Roslin Institute (1997) 'Cloning and Genetic Modification: Use of Cloning in Farm Animal Production' *On Line*, December

42 Stewart, Colin (1997) 'An udder way of making lambs' *Nature*, 27 February
43 Lee, Se-Jin (1998) 'Myostatin is the key to double Muscling' *Ag Biotech* January, Vol 10, No 1
44 Broom, D M (1993) 'The effects of production efficiency on animal welfare' From Proceedings of the Fourth Zodiac Symposium, Biological basis of sustainable animal production, Wageningen, Netherlands, April
45 See ref 20
46 See ref 25
47 *Ag Biotech* Vol 10, No 1, January 1998
48 See ref 23
49 Bauer, S R (1995) 'Food and Drug Administration Concerns Regarding Genetic Improvement of Farm Animals' pp 231–234, Proceedings of Symposium, Beltsville Agricultural Research Center, Beltsville, Maryland, US
50 See ref 49
51 FAC Annual Report 1994; ACNFP Annual Report 1994
52 Hammond, K and Leitch, H W (1995) The FAO Global Program for the Management of Farm Animal Genetic Resources
53 Yu Qing Cao, Mantyh, P W, Carlson, E J, Gillespie, A-M, Epstein, C J and Basbaum, A I 'Primary afferent tachykinins are required to experience moderate to intense pain' *Nature*, Vol 392, 26 March 1998
54 Rexroad, C E Jr (1998) *Why biotechnology? Animal Biotechnology and Ethics*, p 90, Chapman & Hall

14 Genetic Engineering and Food Security

Julie Sheppard

Feeding the world's hungry is what genetic engineering is all about – or so the food industry claims. Poorer countries are often plagued by adverse growing conditions – poor soils, drought, pests and diseases – although many were able to feed themselves until the introduction of cash crops, industrialization and war reduced the available land for growing food. Crops re-engineered to withstand difficult growing conditions, perhaps with built-in resistance to the major pests and diseases, would, claim their proponents, begin to address food security.

Lessons of the Green Revolution?

In the 1970s, similar claims were made about the Green Revolution with its promises to boost yields and increase the efficiency of land use through the introduction of high yielding varieties of staple crops, such as rice. Genetic engineering is billed as the second wave, the Doubly Green Revolution, which will build on the successes of the first.

One crucial lesson gleaned from this experiment was that high yielding hybrid varieties could not be grown in local conditions except by creating an artificial environment, boosting the soil's fertility with fertilizers and protecting against pests and diseases with pesticides. The new hybrids were not adapted to the local geology or climate and could only be sustained by expensive imported chemicals, which poorer farmers could not afford.

So, although the new crops made a contribution to increasing yields for richer farmers, those growing crops simply to feed their families could not afford the chemicals required.

We should remember these lessons today as biotechnology is hailed as the new miracle cure for global food security. The proponents imply that genetic solutions will put food into hungry mouths and redress ecological degradation. But there are important constraints – natural, social and political – on what biotechnology can contribute to feeding the world and potential drawbacks which have been significantly underplayed.

POTENTIAL DRAWBACKS

The potential drawbacks are both environmental and food safety risks:

Genetic overdependency

There are over 220,000 plants species in all, yet only 150 of these are grown commercially. Just 20 provide over 90 per cent of the world's dietary energy. We are already massively over-dependent on a fraction of the species available. This over-reliance on a relatively small number of crops increases the susceptibility of the world's food supplies to pests and diseases. Relying on a limited genetic pool is inherently risky for food supplies across the globe.

The use of GE crops could exacerbate this problem by increasing reliance on a handful of hybrid varieties which, when grown widely, may be vulnerable to attack from pests and diseases. When grown as monocultures, these crops could encourage the establishment of growing systems which are so fragile that plant geneticists will be under constant pressure to develop new varieties to withstand the onslaught of new pests and diseases. Such a genetic treadmill will probably prove as damaging and as self-defeating as the chemical treadmill it is designed to replace.

Increased use and dependence on herbicides

The impact of GE herbicide tolerant crops on herbicide usage is not known and no independent agency will be monitoring the impact. It is possible that they could reduce the overall rates of herbicide use, or at least promote a switch to more environmentally-friendly chemicals.

However, growing plants which are able to withstand increasing doses of herbicide hardly encourages reduced applications. In Australia, for example, new crops have been developed that can tolerate dosages of the herbicide 2,4-D up to eight times those recommended.

Out-crossing: new genes could spread to other plants

There is no way of controlling the transfer of genes for herbicide resistance from crops to 'weedy' relatives, thereby producing 'superweeds' which cannot be easily eradicated. Yesterday's problems could become tomorrow's nightmare, as growers find traditional methods of chemical control no longer work.

The risk of gene transfer to wild species is known to be greater if the new resistant crops are grown in monocultures. Yet these crops have been designed to be grown precisely in this manner, and there are no controls to prevent them being grown in this way.

Weediness: the new crop could become a weed itself

If the newly inserted genes confer a competitive advantage to the new variety, then the crop itself might become a pest, invading the local ecosystem and displacing wild flora. In Britain, some crop plants, for example oilseed rape, have escaped cultivation to become widespread in other areas. It is difficult to predict or control whether any crops modified for herbicide resistance will spread from cultivation to become established in the wider environment.

Resistance

There are two aspects to resistance:

- Resistance to more than one herbicide. Because there is no control over the number or use of herbicide-resistant crops being grown, there is a risk of weeds developing tolerance to not just one but several different herbicides. Multi-resistant weeds could then become impossible to control.

- Bt crops could accelerate the development of resistant pests. *Bacillus thuringiensis* (Bt) is a bacterium which normally lives in the soil and produces a protein which is capable of killing more than 50 species of moth and butterfly. Because of these natural properties, Bt has been used by organic farmers as a biological pesticide for many years.

The Bt gene responsible for conferring a natural resistance to some pests has been inserted into a number of crops. For example, Bt corn is designed to resist the European corn borer, Bt potato to resist the Colorado potato beetle and Bt cotton to resist the pink bollworm, cotton bollworm and tobacco budworm. All three are now grown commercially in the US. There are also field trials of many other Bt food crops currently underway including rice, apple, tomato, aubergine, oilseed rape, alfalfa, walnut and cranberry.[1] The worry is that widespread use of Bt crops may eventually lead to the development of Bt resistance in previously susceptible pests, making its use by organic and even non-organic growers ineffective.[2]

These drawbacks might be seen as a small price to pay for putting food in hungry mouths. But is lack of food or lack of access to the right kinds of food likely to be solved by this latest genetic fix? What are the likely social, economic and political constraints?

CONSTRAINTS

Substitution

Genetic engineering could actually damage the prospects for growing food in the South. These new technologies make it possible for the first time for crops usually grown in the South to be grown in the North. The use of genetic engineering to produce substitutes for raw materials such as sugar, coffee, palm oil and cocoa butter could devastate the economies of the poorer countries which depend on their production. The absence of equitable arrangements to compensate the South for any loss of trade as a result is fuelling fears that biotechnology, rather than their saviour, could be the last nail in their coffin.

Affordability

The commercial drive towards genetic engineering is controlled by a handful of large multinational companies with interests in seeds and agrichemicals. The move to re-engineer crops by linking the sale of seeds with specific herbicides locks growers into commercial packages controlled by particular companies. US growers of Roundup Ready soybean, for example, are being asked to pay an additional $5 a seed-bag as a technology fee and must use a Monsanto approved glyphosate herbicide. It is doubtful if growers in the South can buy into this expensive package, even if it was beneficial for them.

Nutritional impact

According to a 1992 report from the UN, the increased consumption of Green Revolution crops had led in some cases to a falling intake of certain vitamins and minerals. This was because the new crops were low in nutrients and growing them displaced local fruits, vegetables and legumes which were the traditional sources of these nutrients in the diet. It is unclear what impact GE crops will have.

Genetic resources

Future developments in biotechnology will depend on genetic resources found only in the South. Poorer countries are unlikely to share in the benefits (if there are any) for use of these resources and will not be adequately compensated for their safe stewardship of the gene pool on which many of these new crops will depend. Some described the process whereby western companies use the patent system to provide monopolies over genetic material found only in the South as genetic piracy.

THE REAL ISSUES

The contribution of GE organisms to food security will be constrained by the nature of global agribusiness, as Margaret Mellon notes: 'Companies go where the money is, and there is more money to be made in cantaloupe for Americans than in cassava for Africans.'[3]

Moreover,

> *Almost all crop biotechnology research and development is being carried out in the United States, Japan and a few Western European countries, where its impact will be more significant on the development of the biotechnology industry than on food security*

according to Indra K Vasil.[4]

Biotechnology's promise to solve the world's food problems raises false hopes that a mere technological fix can solve a complex economic, social and political malaise.

NOTES

1 *Gene Exchange*, Vol 6, nos 2 and 3, December 1995; and Vol 6, no 4, June 1996
2 'Perils Amidst the Promise: Ecological Risks of Transgenic Crops in a Global Market', Union of Concerned Scientists, Washington DC, 1993; 'Herbicide Tolerant Crops', Department of the Environment/MAFF joint workshop report, London, 1995.
3 *Nature/Biotechnology*, Vol 14, July 1996.
4 *Nature/Biotechnology*, Vol 14, June 1996

15 The Global Contribution of Organic Farming

Patrick Holden

The industrial chapter of global agriculture commenced on a major scale at the end of World War II. The events at that time brought together the common concern about food security with the availability of inputs and technology which could be applied on a large scale to increase yields of staple crops. These were enthusiastically embraced by the governments of most industrialized nations. In the UK, the 1947 Agricultural Act offered farmers guaranteed prices, thus providing them with a bottomless market for everything they could produce. These conditions have persisted for 50 years and, for the last 25, price guarantees have been provided through the CAP operating at an EU level.

However, UK farming is now at a crossroads and there is also a major debate taking place on future strategies for world agriculture. Despite the success in increasing output, the progressive industrialization of agriculture has led to a series of problems which are collectively so serious that there is an urgent need for reform. After 50 years of intensification, landscapes have been irreversibly damaged. Our wildlife is disappearing fast. A succession of food scares has reduced consumer confidence in food quality and safety to an all-time low.

SOME KEY PROBLEMS

Environmental Pollution

Nitrate and pesticide residues are progressively accumulating in our water supplies, so much so that water authorities are now having to spend hundreds of millions of pounds per year removing residues in order to comply with EU safety limits.[1] Pollution today is so bad that water companies are having to strip pesticides out of the drinking water.

Declining biodiversity

Numerous studies by UK conservation organizations have drawn attention to an enormous decline in the biodiversity on UK farms, particularly within the cropped habitat. Walk into a field of wheat in the UK, part the crop canopy and look underneath. You will see nothing but the sown species; the herbicides have seen to that! We have suffered a catastrophic decline in biodiversity in the cropped habitat which is potentially even more serious than the removal of the green fringes and nature reserves, which Dennis Avery (Chapter 2) refers to as 'the wildlands'.

Our landscape has also suffered. The sterile monoculture of wheat and other crops that are grown 'wall to wall', certainly in the eastern counties, have radically transformed our countryside. The hedges have gone. This form of 'cultural memory' that the landscape holds for us all has been taken away from us relatively insidiously during the last 30–40 years.

Soil Quality

The land surface of our planet is clothed in a thin film of organic matter upon which nearly all plant and animal life depends – the soil. A wealth of evidence now exists to suggest that organic matter levels in arable soils have declined to thresholds where increased soil erosion is inevitable. The annual loss of topsoil, one of the most important forms of environmental capital, is now reaching serious proportions. This has been caused by the abandonment of traditional rotations, which maintained organic matter levels, and their replacement with continuous arable cropping.

Food safety

There are pesticides residues in nearly all our food. They may only be present in very small quantities but we have no idea what the cumulative long-term effect is of consuming such residues over a lifetime. The catalogue of recent major food scares – including concern about pesticide residues, especially organo-phosphates; BSE; the rise in incidence of pathogenic organisms, especially salmonella and *E-coli* 0157, which collectively now cause 100,000 cases of food poisoning a year – has now become so serious that the Government has announced the establishment of the Food Standards Agency to deal with them. There is also growing evidence that the resistance to antibiotics building up in bacteria is linked to the prophylactic use of antibiotics in livestock systems.

Food quality

Disturbing new evidence suggests that the use of artificial fertilizers to increase yields also has a negative impact on the mineral and trace element content of staple foods. MAFF statistics on fruit and vegetables over the last 30–40 years show that there has been a consistent decline in the nutritional quality of our food. This is because when high levels of nitrate and phosphate fertilizers are applied to crops it affects the fungi, the bacteria, and other life in the soil. As a result of this one-sided nutrient application there is also a decline in the micronutrient and mineral content and the levels of secondary plant metabolites in our foods. Only recently is this beginning to be measured. A report in the *New Scientist* about South East Asia showed that, after 20 years of Green Revolution techniques, there were alarming declines in the iron and zinc levels. These were linked to a ten point decline in the IQ of the generation of children and to major fertility problems. We may not be seeing that in the UK yet, because most of us have the luxury of a diverse diet, but these problems are there. After all, we are what we eat. The way in which we are fertilizing our crops these days is leading to changes in their nutritional composition. We are literally guinea pigs, as we do not know the long-term effect of consuming nutritionally degraded food over one or more generations.

Animal welfare

The consequences of the industrialization of livestock production on animal welfare are, of course, the major concern of organizations like CIWF. What is beyond question is that the vast majority of UK-produced pigs and poultry live the whole of their lives in conditions which are totally unacceptable from the welfare viewpoint. If animals are permanently housed in sheds, only prophylactic levels of antibiotics stop them becoming ill. There is no choice. These animals also exhibit behavioural tendencies which then lead to cannibalism. The RSPCA's Freedom Food standards allow beak tipping and farrowing crates. I believe this is because they did not tackle the problem of getting the animals out of these intensive conditions. As long as you put animals in conditions where they cannot behave in natural ways, it is inevitable that you will have to constrain the behaviour patterns which result from that confinement.

Wider social impact

Another major impact of industrial agriculture has been a catastrophic decline in rural employment and the destruction of much of the cultural fabric of the rural areas of Britain.

Collectively, these costs of industrial agriculture are enormous, both in terms of their impact on the environment, food safety and quality, human health and the quality of our rural life, although ironically, this impact has yet to be costed and is still not widely understood by many sections of the British public.

In addition, there is a whole generation of scientists mostly signed up to the dominating nature, high-input, hi-tech approach of industrialized agriculture. They still believe that this is the way forward, that we can solve all the above-mentioned problems by more of the same. This is one of the greatest problems that we face today. The fixed minds of those people who still largely control the agricultural industry, the educational and research institutes and who influence government thinking about the way ahead, are the legacy of 50 years of industrial agriculture.

Options for change

Dennis Avery (Chapter 2) and I agree on one thing – that we do have a problem! Clearly it is time for a major change of direction in agricultural policy and practice. The question is, what is the solution?

Two starkly contrasting options are being discussed. One, characterized by Mr Avery, is the 'hi-tech', 'more of the same' solution.

'Food factories and parks' is the first option. At present this is official government policy. It consists of further intensification of the most productive land whilst, in parallel, identifying and conserving the last remaining oases of wilderness and natural habitat. In these areas farmers will be paid to become 'park keepers', receiving payments to undertake specific stewardship activities which help preserve the dwindling wildlife on the margins of their high-tech farms.

The no-holds barred approach in the 'food factories' areas will lead to scientific innovation ruling the day, with GE engineered crops receiving their full complement of fertilizers and pesticides. But:

1 In terms of the environment, food factories will be devoid of all biodiversity, degrade our landscape, continue to pollute water and to erode soil fertility.
2 On the food safety front, since all the recent food scares can be attributed directly or indirectly to intensive farming, there is no reason to believe that if we carry on down this road we will not see further breakdowns which may collectively pose a further threat to public health.

I suggest that feeding the world with plastics and pesticides is not a satisfactory solution, partly because it is unsustainable and also because it fails to meet the requirements of today's consumer in a number of very specific ways.

Current orthodoxies need challenging. I believe there could be another, better way forward. This approach is best defined as organic farming. For over 50 years the Soil Association has promoted sustainable agriculture – organic farming – which combines the production of safe, high quality food with farming practices that protect the environment.

Organic farming reduces external inputs and maximizes internal nutrient cycles. It uses high levels of husbandry skills, high management intensity, instead of replacing these with inputs, which is the old system. We have been using inputs to produce and control problems. The challenge is to find management and husbandry solutions to the problems and maintain the health of the system rather than having to treat the ill-health consequences of the results. The organic solution promotes biodiversity, through crop rotations. The integrated approach is not a low input, low output approach, as Dennis Avery suggests. It is high management input and a potentially high output approach as well. In my view there is no reason why the majority of UK agriculture should not be switched to organic production over the next 20 years.

ORGANIC PRINCIPLES

The characteristics of the integrated farming system I propose are best defined in the organic standards. These were developed over the past 25 years by the Soil Association and other organizations all over the world as a prescription for the application of sustainable agriculture principles.

The characteristics of organic farming include food quality, nutrient cycling, biodiversity, integration with the environment, allowing crops to coexist with non-sown species, high levels of animal welfare, getting the animals out of sheds and onto fields, reducing their proneness to disease, cultural benefits through local food links, less centralized production systems, more employment and closer relationships with the environment. These are the characteristics of organic food production systems and they have the potential to become the predominant form of agriculture of the future.

Within those principles, there is no place for genetic engineering. The Soil Association believes that there is no place for genetic engineering in agriculture as a whole. We see the technology as merely an extension of current high input farming, substituting gene splicing for pesticides. Genetic engineering is a major threat to the environment, poses potential risks to human health, removes the right of farmer and consumer choice and is incompatible with the principles of sustainable agriculture.

Dennis Avery and I cannot both be right. The Soil Association believes Mr Avery is wrong in three ways:

- Firstly, he is conceptually wrong because the high input approach is incompatible with the principles of sustainable agriculture. Monocultures are inherently unstable. In a monoculture, unwanted things will happen – salmonella will infect livestock, or weeds will invade cereal crops. Nature abhors a monoculture. We must challenge the premise that only the sown crop is the desirable outcome. That is the direction that biotech takes – breeding crops, for example, with herbicide resistance, so the aim is to get the crop and nothing else in the most intensively cropped areas. That system will be inherently vulnerable to various forms of outside threat including fungus, pest and weed problems. We do not always know what the threats are until they happen, but there will be breakdowns. It is the basic approach that is conceptually flawed.

- Secondly, the idea of saving the wildlands in one area while increasing farming intensity in others is also conceptually flawed. Nature conservation and the preservation of biodiversity should be integral to sustainable agriculture. Then we do not get into this terrible conflict where we feel we have to turn most of our productive farmland into factories and ignore nature there, and have other places with some sort of artificial remnants of the natural world.
- Thirdly, this intensification is not necessary to feed the growing world population as Mr Avery suggests. Whatever the case may be with the one organic farmer he talked to in Britain, his claim that organic farming is achieving only 50 per cent of the yields of conventional agriculture is simply not true. It is certainly not true on my organic farm. Organic farmers generally, when they are working well with effective farming systems, can equal and in some cases exceed conventional yields because they build soil fertility. They are developing a harmonious relationship with nature and that can actually be regenerative, not exploitative.

For instance, until recently, pigs were not basically grain consumers, they did not consume the product of vast prairie acreages of cereals. Instead, they ate products which otherwise would have been either wasted or were grown in relatively natural systems. We have turned the millions of pigs in intensive farms into cereal eaters – competing directly with humans. There is no reason why we cannot produce livestock *not* relying primarily on food which would otherwise be consumed by human beings. That is the key with both ruminants and non-ruminants and it is a very important flaw in Mr Avery's argument.

Jules Pretty has done a great deal of research on organic and near organic sustainable agricultural systems around the world. His research shows that millions of farmers in South America, in Asia and in Africa, are moving towards more sustainable lower input, higher management systems and are achieving remarkable yields. These yields are being achieved not only on the lands which are inherently suitable for agriculture but also in less favourable environments. In integrated food productions systems in relatively unproductive areas they are making them remarkably sustainably productive.

THE POTENTIAL OF ORGANIC FARMING

As an organic farmer for 25 years, I find when these principles are put into practice I am constantly amazed by the possibilities for improvement. If we devoted the sort of research and development resources to sustainable organic farming systems as have gone on industrialized farming systems, I believe that we could produce enormous developments using conventional plant breeding and the like.

In future, the International Federation of Organic Agriculture Movements (IFOAM) will have a growing role to play. If we have to keep the WTO, and we probably do, then we need criteria for sustainable trade and the standards and principles of IFOAM should be those criteria. It may sound like a pipe dream, but I do not think we have any right to trade food internationally unless the production methods are in line with the principles of sustainability. They should be the new criteria for sustainable trade

Within the EU, there is now a real possibility for fundamental reform of the CAP. We need to abandon the present structures and support schemes and replace them with area payments which should be conditional on sound farming practice including environmental protection. Some officials within the Commission believe that would not be possible to administer, but there is a growing lobby trying to influence reform of the CAP and to progressively move to an area payment system.

The UK still has a very backward policy towards organic farming but we have made some progress. The existing organic aid scheme is based on area payments and is currently being reviewed. New features include increasing the levels of conversion supports from £250 to £450 per hectare and there are a number of other minor changes which will make conversion a more attractive option for new entrants. But it is nowhere near far enough. There are two problems. One is that MAFF's vision for organic agriculture seems to be as a niche market, not as a major strategic option for the agriculture of the future. That is also reflected in MAFF's view of organic farming as a means of protecting the environment. Organic farming has tremendous potential to protect the environment on a wide scale. On the 85 per cent of land which is not covered by various stewardship agreements, organic farming could protect birdlife and other biodiversity.

Organic farming potentially could, and I believe eventually will, become the predominant form of sustainable agriculture in the early part of the 21st century. If we are going to feed the world, sustainable solutions are the only way ahead. The high input approach does not work. It will continue to create many more problems similar to those experienced recently both in food safety and quality and environmental damage.

One final point. Perhaps an accurate indicator of a civilized society is the way in which it treats old people and its animals. The abuse of animals that has been part of the industrialization of agriculture over the last 30–40 years has incurred an enormous moral debt. If we are going to repay that debt, a sustainable agriculture as defined in organic standards, which incorporates at its very heart high levels of animal welfare, is the best way to start.

Notes

1 Soil Association (1996) *Counting the Cost of Industrial Agriculture*, Bristol

16 Beyond the Killing Fields: Working Towards a Vegetarian Future

Mark Gold

World meat production quadrupled from 44 million tonnes in 1950 to 195 million tonnes in 1996,[1] largely to feed the populations of wealthier nations like the UK. This expansion was only made possible by the intensification of livestock farming. It allowed vast numbers of animals to be confined in small areas and increased their productivity through the ruthless application of genetic science, pharmaceuticals and carefully controlled feed regimes.

In the developed world, exploitation of animals is at the heart of food production. The accepted wisdom is to devote the majority of agricultural land to grow feed for livestock which is then processed by them into food for humans. It is a hopelessly inefficient system. Far greater quantities of food could be obtained by growing plants for direct human consumption, but, as supermarket shelves groaning with seemingly infinite choices testify, problems of waste and inefficiency are easily masked in a land of apparent plenty.

In contrast, the FAO estimates that approximately 800 million people in developing nations are severely malnourished.[2] Some of this suffering results from conflict, inept and corrupt politics or natural disasters, but even so, millions of humans die from lack of food while – theoretically at least – there is more than enough produced worldwide to ensure that nobody needs to go hungry.

Throughout the 20th century, economic development and increased meat production have consistently gone hand in hand. It is hardly surprising, then, that poor people and countries try to emulate the consumption patterns of the rich. Regardless of food shortages, in all but the most poverty stricken areas of the developing world, people are following the fashion of richer nations by increasing their dependence upon food derived from animals. This is good news for the meat industries of the developed world – enabling them to profit by providing all the paraphernalia of intensive production such as special feeds and pharmaceuticals, technology, highly productive genetic breeds, consultancies and so on. But as a policy to help hungry people, it is an inevitable disaster.

POULTRY AND PORK

The poultry industry illustrates the point. Between the 1960s and 1990, consumption of poultry meat increased dramatically in almost every country, with a worldwide average rise of 50 per cent per person.[3] This massive growth created business opportunities for Western factory farming companies all over the world. When *Poultry World* magazine highlighted 'the great scope for expansion' in Africa, it emphasized how African countries are 'totally dependent for commercial table birds and egg production on directly imported stock'; that the 'feed industry is dominated by two or three large commercial feed compounders ...based in Europe and America who own franchised industries'; and that African poultry farmers are also 'dependent on the importation of a large group of pharmaceuticals'.[4] It is a similar story in many parts of Asia, too.

Wherever the poultry industry goes it takes with it the same problem. How do countries struggling to produce enough grain to feed people find chicken feed? It takes about 2.8kg of grain to produce only 1kg of poultry,[5] so the only choice is either to divert local agricultural land from the production of human food, or else to import grain to feed to the birds. But who eats them after they have been fattened? It is not those most in need, because they could not even afford the relatively low-priced grain used to produce relatively expensive meat.

The poultry industry is well aware of its own limitations. In India, the broiler chicken industry is expected to double to 820 million birds per annum by the year 2000 (again many of the birds supplied by western multinationals). The Indian branch president of the World Poultry Science Association, Anuradha J Desai, categorizes the target audience as 'the fast growing middle class of over 250 million potential customers'.[6] No mention of the poor.

If and when the financially secure middle classes get to feast upon increased quantities of poultry meat, the inevitable consequence is less grain available for those people in desperate need of any food. Meat eating is not only a potent symbol of the massive inequalities of food between rich and poor areas of the planet, but is also creating widening gaps between rich and poor locally.

The problem is exacerbated because many developing nations such as China, Indonesia, Iran, Pakistan, Nigeria, Ethiopia, Mexico and Egypt are already net importers of grains.[7] More meat production (not just poultry) adds inevitably to an increase in demand and extra pressure on the international harvest. A vicious circle is perpetuated. Grain becomes scarce; prices rise. The poorest nations of all cannot afford to compete. Less food is available; more malnutrition follows. Bizarrely, sometimes countries facing economic ruin raise capital by selling grain to richer countries as animal feed – Ethiopia sold crops to the UK at the height of its 1984 famine and North Korea, suffering desperate food shortages in 1997, exported 1,000 tonnes of maize to fuel the growing Japanese poultry market.[8] Less dramatically, thousands of hectares of land in developing countries are used to produce animal feed for richer nations ever year.

The destructive consequences of the poultry industry are mirrored by expansion in other areas of the meat trade. In China, for example, an astronomical rise in pork consumption over the last decade means the Chinese now consume more per person than in the meat-obsessed US.[9] This has transformed China from being an exporter of 8 million tonnes of grain in 1993 to becoming a net importer of 16 million tonnes by 1995 because pig and other forms of animal production are even less efficient in the converting of grain to produce meat.[10]

FEED VS FOOD

If China and other developing countries continue to demand more animal produce we will face a world food problem of catastrophic proportions because the world's grain harvest cannot keep expanding to meet the needs of both a rapidly growing human population and an explosion in the number of farm animals. The pressure is already beginning to tell. Throughout the 1990s there has been a steady reduction in the amount of food theoretically available for each person. The world's grain harvest is failing to expand sufficiently to nourish 90 million extra human mouths which have to be fed each year. The more meat consumed, the steeper the decline in available food per person will become. If everyone were to imitate the habits of the US – where it takes 800kg of grain per person to sustain a diet in which 25 per cent of calories are obtained from animal produce – then the planet could sustain roughly only 3 billion people. The current world population is already 5.7 billion, and is forecast to rise to 11–14 billion within 40 years. On a near vegetarian diet, however, about 200kg can provide all the calories an adult needs in a year.[11]

The same wasteful inefficiency applies equally to water consumption. Ground water and surface water are increasingly scarce and 'saving water is becoming as important to human survival as saving energy'.[12] Water tables are falling drastically in many nations, including parts of India, China and the US.[13] The growing farm animal population adds insatiably to the demand for water. A kilo of beef protein takes up to 15 times more water to produce than the equivalent value of vegetable protein; a kilo of chicken meat takes nearly twice as much as a kilo of rice or soya beans.[14] Vegetarians need less than a third as much water to sustain their diet as meat eaters,[15] making vegetarianism not only the diet for a small planet, but also the only rational response to a dry one.

Intensive animal farming also creates other environmental hazards. It is a major source of water pollution. 'Cattle and other livestock account for twice the amount of pollutants as come from all US industrial sources' according to Jeremy Rifkin;[16] cattle and sheep are also among the biggest sources of methane, a gas second only to carbon dioxide in its contribution to global warming. Throughout the world, livestock herds accelerate erosion and desertification with, for example, 85 per cent of topsoil loss in the US attributable to ranching.[17] And 'Hamburgerization' – the felling of forests for cattle ranching – is one of the chief causes of rainforest destruction. Since 1970, 'farmers and ranchers have converted more than 20 million hectares of Latin America's moist tropical forests to cattle pasture'.[18] In all, Central America has lost more than one third of its forests since the 1960s, while land in pasture has increased by at least 50 per cent.[19]

Meat probably also damages many of those who consume it. Ironically, given increased meat consumption in China, it was there that the largest ever epidemiological study into the relationship between diet and health assessed 6500 people across 65 Chinese counties between 1983–88. China had been deliberately chosen as the country offering the best possible dietary contrasts for research purposes.

The project coordinator was Professor Colin Campbell, former administrator of human nutrition for the USDA. The conclusion was unequivocal: 'we're basically a vegetarian species and should be eating a wide variety of plant foods and minimizing our intake of animal foods'.[20] The results showed clearly that:

> *the closer a region's diet came to the Western model – high in animal protein and fat, low in fibre – the more the people suffered from the so-called diseases of affluence that kill people in the US and other developed countries – various cancers, leukaemia, diabetes and coronary heart diseases.*[21]

Choosing a Future

Carry on regardless

How can we meet the growing threat to our capacity to feed an expanding human population? The current political choice is largely to maintain seemingly undaunted faith in the power of science and technology – financed largely by big business interests – to find solutions which will allow us to continue with similar policies to those pursued relentlessly in the last half a century. Grain production has, it is argued, regularly increased to meet rises in population, despite frequent warnings that it would not manage to do so. Why should this not continue to be the case? With more efficient use of fertilizers and other agricultural chemicals, new technologies to expand yields and more land brought under production, it is contested that it will still be possible to meet the needs of both a growing human population and a similarly escalating number of farm animals bred for meat.

But will it? 'In country after country, farmers have discovered that they are already using the maximum amount of fertiliser that existing crop varieties can effectively use,' says Lester Brown, President of the Worldwatch Institute.[22] In other words, even if we ignore the destructive environmental impact of the current range of agricultural chemicals, their capacity to expand harvests further is now negligible. Moreover, the amount of extra land available to be brought into cultivation for conventional agriculture is also severely limited. In fact, in many places potential food producing areas are being lost, both to other development and to soil erosion caused by intensive methods. This leaves GE crops as 'the great white hope' for propping up current policies. The multinational investors in the new biotechnology make massive claims for the capacity of their new products both to increase food supplies and to reduce chemical inputs, but even if this proves to be true (and there are genuine doubts), at what cost? The potentially lethal effects of herbicide-resistant genes being transferred to non-target plants are daunting. (See Sheppard, Chapter 14.) So too are the political implications of allowing the few massive biotechnology companies that control the new GE crops such immense power over world food production.[23]

Organic alternative

Opposing this 'carry on regardless' mentality is the considerably more enlightened philosophy of the organic movement. Briefly, its message is:

- conquer our dependence upon non-renewable resources and chemicals which damage eco-systems;
- minimize waste and inefficiency and return all organic waste to the soil to enhance fertility.

As for animals, the organic movement generally encourages moves away from a predominantly meat-centred diet, but argues that some dependence upon them is desirable. Organic farmers agree the most productive farmland is best turned over to plant foods for direct human consumption, but see livestock as 'an essential part of the organic cycle' on marginal lands and uplands.[24] Where crops cannot easily be grown, animals can convert grass into food, helping to conserve the existing environment, to produce food where no plants could otherwise grow and to yield valuable manure to help maintain soil fertility. Animal welfare is seen as important. Providing decent living conditions and a reasonable length of life before slaughter are integral to the system, but ultimately, killing for meat, blood and bone meal is also considered essential. Some organic growers are actively hostile to the idea of an entirely plant-based agriculture, arguing that 'a rotation including grass and livestock is essential'.[25]

UNJUSTIFIED KILLING

Like many vegetarians, I have some reservations about organic mixed farming. I cannot accept, without reservation, any philosophy of food production which still depends upon the slaughter of animals at a relatively young age so that humans can feed on their dead bodies. I question the acceptance in organic mixed farming of slaughter as, in some way, natural and necessary. Nevertheless, I also recognize it as a practical way forward from the current ethos, worthy of enthusiastic support provided that is not presented as the ultimate in non-violent agriculture.

I believe that we can no longer justify the killing of animals for food on the grounds that it is their 'natural' role. Every cow, sheep, pig, chicken, goat and rabbit is an individual, capable of experiencing pain and – to varying degrees – some of the emotions we like to think unique to humans. As the great American humanitarian John Howard Moore expressed it nearly a hundred years ago:

> *man is not the pedestalled individual pictured by his imagination...He is a pain-shunning, pleasure-seeking, death-dreading organism, differing in particulars, but not in kind, from the pain-shunning, pleasure seeking, death-dreading organisms below and around him.*[26]

In our hearts we know this to be true. We also know that slaughter – however it is dressed up in such terms as 'humane' – is repugnant. Nothing illustrated this more forcefully in Britain than when two Tamworth pigs, Butch and Sundance, escaped from the back of a lorry at a Wiltshire slaughterhouse in spring 1998, swam across an ice-cold river and then avoided recapture for over a week. It is scarcely an exaggeration to say that their fate gripped the nation. Acting 'on behalf of its readers', a national newspaper eventually paid to ensure that the two escapees were saved from slaughter, guaranteeing that they would live out the rest of their lives in a local sanctuary.[27] It was a happy ending, enthusiastically supported by the vast majority of public opinion.

The incident could be dismissed as an example of acute sentimentality when thousands of other pigs are killed routinely for the dinner table each week. More positively, it provides evidence of a profound recognition of animal sensibility and intelligence lurking near the surface of everyday human consciousness. Despite the often huge gap between our deepest feelings and our actions, there is an evolving acceptance of the complexity and sensitivity of all animals which must eventually find expression in a reevaluation of the millions of deaths we take so carelessly for granted.

A plant-based food system

Even if everyone agreed that animal slaughter is distasteful, is it not necessary? Would it really be possible to create a system of food production based upon a philosophy of non-violence to other creatures? Is an exclusively plant-based system feasible? Can it provide enough food? Can it maintain the health of the human population? Is it possible to maintain soil fertility without animal manure? Would the unique character of the environment be destroyed without livestock farming? These are among the main issues which need to be addressed for a truly compassionate farming system.

Could enough food be produced?

The lower down that humans eat on the food chain, the less land it takes to sustain us. One estimate suggests that an adult can be sustained on about one fifth of an acre on a vegan diet – far less land than for any other systems of food production.[28] In the UK, this would mean that the whole population could be fed on roughly 10 million acres. There are currently about 40–45 million acres of agricultural land available.[29]

Plant-based diets also lend themselves readily to small-scale production, allowing more people to be involved in maximizing the amount of food grown locally for local need. According to Argentinian specialist Pablo Gutman, in Buenos Aires a plot of 10 metres square intensively cultivated could produce all the vegetables needed by a family of five.[30] In an ideal world, this would mean smaller communities striving towards near self-sufficiency.

Ironically, it is contemporary initiatives in heavily populated urban areas that give a glimpse of the potential for non-animal systems of food production. An estimated 200 million urban dwellers worldwide provide 800 million people with at least some of their food requirements. In Kampala, Uganda, for example, 35 per cent of households produce their own food, including cassava, sweet potatoes, coco yams, plantains, maize and beans. The possibilities of using similar small-scale production to combat human hunger are severely undervalued. Up to 50 per cent of the total area in many cities in developing countries is vacant public land, at least some of which could be reclaimed for food production. A 'small is beautiful' approach could also make an equally relevant contribution in richer countries of the world. Reclamation of waste land for vegetable production, for example, made a crucial impact on the wartime economies of both the US and the UK, with US 'Victory Gardens' yielding 44 per cent of all the fresh vegetables produced in the US in 1944.

Although abundant amounts of additional food could be produced by switching to a plant-based diet, animals do not always compete with humans for food. Pigs and poultry fed on scraps and ruminants grazing on uncultivated grassland are obvious exceptions. Fish, too, are living creatures which provide food where nothing more productive could be obtained. The key point, however, is that it is theoretically possible to grow more than enough food exclusively from plant sources to ensure that there would not be any need for slaughter.

Would veganic agriculture turn the countryside into a vast, featureless prairie?

Abandoning livestock production need not leave the kind of arable farming practised today with huge fields sown to grains and trees and hedges removed to allow easy harvesting by large machinery. The aim would be to maximize the use of small plots on which to grow vegetables, enclosed by mixed hedgerows. Cereals and protein crops would be cultivated in larger fields (though smaller than the average size today), surrounded by food-bearing trees.

Food could still be grown in many upland and marginal areas currently abandoned as unfit for cultivation. Israel, for example, created fertile areas from desert after World War II, and the inspiring vision of the late Sir Richard Barbe Baker helped to transform a sandy coastline into the gardens of the Findhorn Community and also motivated the reclamation of many square miles of the Sahara. Sir Richard believed, and to a large extent demonstrated, that 'with trees and proper care, good soil can, in due course, be produced practically anywhere'.[31] The key, he believed, was well-managed forestry.

A vegetarian world would need to be tree-centred. There are few areas in the world where some food-producing trees will not flourish, able to provide vast quantities of health-giving and energy-efficient foods. According to regional variations, mixed forests containing different leguminous, nut and fruit trees could provide food in greater quantities and far more efficiently than animal farming. Many nuts and legumes far outweigh the potential of any other foods both in yield per hectare and nutrient value.[32] Priority should also be given to the development of leaf protein, using the food potential of many types of tree leaves. 'Leaf soup' – a curd obtained using simple technology – has already been used in some developing countries, offering a 'greater yield of protein per acre than any other crop'.[33] Schemes to use it as a basis for a range of more sophisticated products are also underway in richer areas of the world. If only a fraction of the vast technological effort presently employed producing more profitable 'novel' convenience foods was devoted to the development of tasty and healthy products based on a rational use of resources it could open up enormous possibilities.

In forest regeneration, aspects of permaculture could be introduced into mainstream production by using different layers of the canopy to maximize harvests without damaging the environment – tall trees planted alongside smaller trees, large shrubs, herbs, ground layers and plants and climbers all intermingled.

Apart from producing enormous amounts of food, a tree-based culture would have major environmental benefits. Vast quantities of carbon dioxide would be absorbed by leaves, reducing the impact of global warming; soil erosion would be checked and the threat from both drought and flood diminished by the capacity of trees to soak up heavy rainfall through their leaves and allow moisture to seep slowly back into the soil. 'By growing enough trees, we can satisfy nearly every human need, including that for food, and at the same time do much to restore and maintain planetary health,' argues Kathleen Jannaway.[34]

Could soil fertility be maintained without animal manure?

Already, many examples of small-scale, successful organic plant food production without the use of animal products exist. Many are simply the result of circumstances – on some urban garden plots or roof gardens where no animal manure is available, for example. Others are pioneering projects consciously promoting a vegetarian future, such as the Khadigar farming community in Farmingham, US. Here vegan organic methods are applied on a 95-acre site to grow vegetables, oilseeds, grains, soya beans and other food legumes, plus soft fruits, green manures, flowers, herbs and flax.[35]

Apart from green composting techniques, a possible vegetarian future requires replacing animal manure with human wastes. This turns the problem of sewage disposal into a vital asset in crop production. Without treatment, human effluent is a health risk, creating disease-producing bacteria, but the techniques to overcome such problems are already well-established. The aerated-static piles method developed by the USDA in the 1970s has 'revolutionized municipal recycling of organic wastes' according to the Worldwatch Institute. Several schemes are now using human effluent successfully. By 1991, Shanghai was collecting 90 per cent – some 8000 tonnes per day – of human waste from its inhabitants.[36] After treatment, the compost produced is sold to local farmers. On a small, local scale, the innovative Centre For Alternative Technology in Wales treats all waste and reuses it on its flower and food beds.

Would plant-based agriculture destroy the established environment?

Eliminating animals from the food chain would lead to radical changes in land use. It might affect biodiversity in some areas but it should ultimately be an enriching influence. On a global scale, livestock farming threatens flora and fauna rather than preserves it. In Africa, for example, overgrazing has caused massive desertification and the loss of many species of wild fauna.[37] On a less destructive local scale, overstocking of thousands of heavily subsidized sheep has created what FoE describe as an 'ecological disaster' at one of the Three Peaks in Yorkshire. Damage over several thousands of acres has resulted in the destruction of several rare plant species.[38]

Veganic agriculture would result in an evolution of the landscape, rather than its destruction. Concern for preserving diversity of species should not be confused with a conservative desire to oppose all change. A world that abandons the exploitation of animals on ethical grounds would intrinsically be more compassionate and caring, with environmental protection for both practical and aesthetic reasons a top priority. Conservation of the landscape and wildlife is integral to the vision of plant-based agriculture. For those traditionalists who would like to see areas of uplands maintained by grazing animals, some land could be set aside where some animals could be looked after on well-managed sanctuaries, without facing the betrayal of the abattoir. Enough land should be available for this.

Can the human population be fed healthily on an exclusively plant-based diet?

Yes. The evidence is overwhelming. Not only do third generation vegan children prove the point, but so too does a good deal of independent nutritional research. There may even be considerable health advantages associated with a vegan diet. 'Lacto-ovo, lacto-vegetarian and vegan diets have decreased incidence of cancers in general' and a wide range of plant foods contain 'an extensive variety of potential cancer-preventive substances', according to the comprehensive WCRF, *Food, Nutrition and the Prevention of Cancer: a Global Perspective* (See Cannon, Chapter 11).[39]

Vegetarian and vegan diets may include an enormous range of different foods. Critics often imply that a plant-based diet is inevitably bland and monotonous – yet there are more than 6,000 known edible plants in the world.[40] Many vegetarians and vegans eat a considerably more varied diet than some meat eaters.

For health, a good vegan diet must mean a varied diet. While food policies should always be centred upon the principle that as much food as possible should be produced locally for local need, fair trade between nations would allow for the import and export of food plants which local growing conditions exclude.

A RADICAL RETHINK

To suggest an idealistic world, far removed from the current ethos, has both advantages and disadvantages. It is easy for critics to dismiss the vision as detached from the blood, sweat and grime of everyday living, but not even those who believe the ideas to be utter nonsense can prove conclusively that their judgement is correct! Inevitably, the above vision of a vegetarian future is a simplification which ignores the largely unpredictable problems that any unestablished system of food production might create. The cultivation of food crops depends upon living organisms and climatic variations, which ensure that no proposed system can ever be completely disaster-proof or applicable to every society. Furthermore, broad terms such as 'vegetarian diets', 'world food production', and 'meat' ignore factors such as quality and the vastly varied political and geographical conditions which exist around the world. Certainly, the diet forced on some of the estimated one billion vegetarians worldwide by poverty is as unhealthy as the junk-food meat diet of the poorer members of developed nations.

A perfect global system of harmony is unachievable. Nonetheless, radical change can be achieved. To prove the point, imagine it. The year is 1798. Interested parties have gathered to discuss how human beings will organize and feed themselves 200 years in the future. One speaker rises and presents the idea that the best solution will be to rip out hedgerows and trees and then use vast machines and chemicals to grow grains in massive prairie-like fields. These will then be gathered and fed to millions of 'animal machines' kept closely confined in environmentally-controlled windowless sheds. After the animals have been fattened they will be transported in large numbers by huge, mechanized vehicles to slaughterhouses capable of killing thousands daily. Indeed – our imaginary spokesperson claims – in the UK alone by the year 2000 we will be slaughtering two million chickens every day and none of them will ever have felt fresh air or seen sunlight until the final journey to have their throats cut. This is the system we have now. Yet to have predicted such methods 200 years ago would have been near impossible – very few could have possessed such a grotesque imagination.

As bizarre and unachievable as my vegan vision may seem to at least some readers, it is no more so than the practices of today would have been to past generations. Moreover, I want to emphasize this: is cruelty-free food production at least worth working towards? Compare it to what we have now. Would the world really be less civilized under a system based upon a first principle of plant-based food grown locally for local consumption through small-scale community schemes? Is it really so bizarre to argue that we should treat all living creatures with a kindly heart, inflicting as little pain and misery upon them as we would wish to suffer ourselves? Or that we should prioritize the regeneration of forests and turn our human wastes into a source of fertility rather than pollution?

My argument is based upon rational and practical ideas. It is not science fiction or fantasy. It is theory, based upon existing human experience, albeit on a relatively small scale. If only the vast resource of human intelligence and ingenuity that is wasted on projects such as developing weapons of mass destruction were instead focused passionately upon developing a compassionate and rational food-producing policy, there is no reason why the world could not move rapidly towards a meat-free agriculture.

Scientific ingenuity or political theory will not define our ultimate progress. That can only be measured by the degree to which the compassionate elements of our character are able to flourish at the expense of the destructive greed which plagues us in so many ways. This is why those environmentalists who consider vegetarianism and animal protection an irritating distraction are mistaken. For whatever crises humanity creates for itself in the next millennium to follow the current problems – globalization, global warming, genetic engineering of crops and so on – they will not be solved by technical expertise alone. As the great humanitarian Henry Salt said nearly 100 years ago:

> *The cause of each of all the evils that afflicts the world is the same – the general lack of humanity, the lack of the knowledge that all sentient life is akin, and that he*

> *who injures a fellow-being is in fact doing injury to himself... As long as man kills the lower races for food or sport, he will be ready to kill his own race for enmity. It is not this bloodshed, or that bloodshed, that must cease, but all needless bloodshed – all wanton infliction of pain or death upon our fellow-beings.*[41]

I do not know whether we will ever achieve a vegetarian human world. But I believe we need desperately to pursue a civilization where the convenience of chemical poisons and factory farms is abandoned in favour of a conviction that every difficulty which arises should be confronted by the least violent and least wasteful response available to us. Realistically, it is not always possible to avoid killing – if 'pests' invade your crops and all preventative methods fail, for instance – and sometimes humans do have to struggle for their own space in the natural world as does every living creature. Nevertheless, a belief in not slaughtering animals simply to satisfy our own taste buds is part of a deep-rooted challenge to the many unenlightened and brutal practices which still characterize many aspects of twentieth century living. Perhaps future generations may yet embrace the spiritual truth of Leo Tolstoy's assertion that we have failed so dismally to grasp: 'as long as there are slaughterhouses there will be battlefields'?

NOTES

1 Figures obtained from the Worldwatch Institute
2 WCRF/AICR (1997) *Food, Nutrition and the Prevention of Cancer: a global perspective*
3 FAO, 1996
4 *World Poultry*, February 1989
5 Alan B Durning and Holly B Brough, *Taking Stock*, Worldwatch Institute, Paper 103, July 1991
6 Dr M A Ibrahim, 'Great Prospects For The Indian Poultry Industry', *World Poultry*, Vol 12, No 7, 1996
7 Lester R Brown, *Facing Food Scarcity*, Worldwatch, November/December 1995
8 Reported in *The Independent*, 18 July 1997
9 Lester R Brown, *Who Will Feed China?* Worldwatch, September/October 1994
10 See ref 7
11 Statistics in this paragraph from Brown, ref 7
12 Fred Pearce, 'Thirsty meals that suck the world dry', *New Scientist*, 1 February 1997

13 See ref 7
14 See ref 12
15 Jeremy Rifkin (1992) *Beyond Beef*, Penguin Books
16 Ibid
17 Joni Seager (1995) *The State of the Environment Atlas*, Penguin Books
18 Alan B Durning and Holly B Brough, *Taking Stock*, Worldwatch Paper No 103, July 1991
19 Ibid
20 Sharon Boyd-Peshkin (1992) 'Counting On China' *Vegetarian Times* (US), April
21 'Huge Study of Diet Indicts Fat and Meat' (1990) *New York Times*, 8 May
22 See ref 7
23 See, for example, 'Biotech firm has eyes on all you can eat', George Monbiot, John Harvey, Mark Milner and John Vidal, *The Guardian*, 15 December 1997 and 'The risks are not understood. And the livelihoods of millions of people in the Third World are threatened', Vandana Shiva, *The Guardian*, 18 December 1997
24 Richard Young 'Livestock – an essential part of the organic cycle' *Living Earth*, The Soil Association, No 197 Jan–March 1998
25 Ibid
26 J Howard Moore (1992) *The Universal Kinship*, Ed Charles Magel, The Kinship Library, Centaur Press
27 Nationwide, BBC Radio 5 Live, 3 February 1998
28 Kathleen Jannaway (1991) 'Abundant Living In The Coming Age Of The Tree', Movement for Compassionate Living, March
29 Ibid
30 Statistics in this paragraph from *Closing The Nutrient Loop*, Toni Nelson, Worldwatch, November/December 1996.
31 Quoted from 'A New World Order', Kathleen Jannaway, Movement for Compassionate Living, 1997
32 See ref 28
33 'Feeding The World', *New Scientist*, 30 July 1987
34 See ref 32
35 Will Bonsall, 'The Khadighar Community – Ethical Farming In Action' *Vohan News*, Issue 1, Vegan Organic Horticultural Agricultural Network, May 1997
36 Toni Nelson (1996) *Closing The Nutrient Gap*, Worldwatch, November/December 1996
37 See ref 15
38 'Hill sheep put rare plants at risk', Martin Wainwright, *The Guardian*, 10 June 1998
39 See ref 2
40 Alan Wakeman and Gordon Baskerville (1986) *The Vegan Cookbook*, Faber & Faber
41 Henry Salt (1921) *Seventy Years Among Savages*, George Allen and Unwin

Part VI
Trade, Welfare and Values

Food is a global business – what happens in one place in our world can soon affect very different localities. New trade rules and the whole globalization of economic life are having a huge impact on people's, animals' and the planet's well-being – as well as on the wealth being accumulated by the few. But what is this all for; what values does it serve? Do we need to get down to the heart of things if we are to develop a trading system that really meets people's needs? These are the issues discussed in Chapters 17–21.

17 Trade Rules, Animal Welfare and the European Union

Peter Stevenson

The GATT – the General Agreement on Tariffs and Trade – dates back to 1947. Only recently, however, has it begun to strike terror into the hearts of those concerned with animal welfare. In 1994, the 'Agreement establishing the World Trade Organization' was adopted. This did not amend the main GATT Articles but it introduced a number of highly significant and – for animal welfare – damaging changes:

- It includes a large number of side agreements that were negotiated in the Uruguay Round. Of these, the Agreement on Sanitary and Phytosanitary Measures, the Agreement on Technical Barriers to Trade (TBT) and the Agreement on Agriculture are important for animal protection – although the Agreement on Trade-Related Aspects of Intellectual Property Rights (TRIPs) will also have effects.
- It established the WTO – based in Geneva. Its role is to facilitate free trade and ensure compliance with the provisions of GATT 1994 and its accompanying side agreements.
- Crucially, a new approach to settling disputes was agreed. Under GATT 1947, countries could, in effect, largely ignore the report of a dispute panel. Under the 1994 WTO Agreement all that has changed. Now panel reports are binding on the parties unless every member country, including the winning party, decides not to adopt the report. Where a country fails to comply with a panel ruling within a reasonable time, either it has to pay compensation or the winning country is entitled to seek permission from the

Dispute Settlement Board to introduce retaliatory trade measures against the losers. In short, the GATT no longer fires blanks.

It is the EU, and not its individual member states, which is a member of the WTO. Already GATT 1994 is having a disastrous effect on animal welfare. Two EU measures agreed in the early 1990s – the Leghold Trap Regulation and the Cosmetics Directive – have largely been unravelled because the EU feared that they would fail to survive WTO challenges.

Worse still, the danger is that the EU will back away from introducing new animal protection laws because of fears that they are inconsistent with WTO rules. It may, for example, shy away from extending the moratorium on the use of BST in the EU when it expires in 1999 or banning the battery cage.

Several key GATT provisions threaten the introduction of improved welfare standards:

- Article XI prohibits bans or restrictions on imports and exports. It in effect states that:

  > *no prohibitions or restrictions shall be instituted or maintained by any contracting party on (i) the importation of any product from any other contracting party or (ii) the exportation of any product to any other contracting party.*

  Article XI prohibits not just import bans but also export bans. Each year dealers export some 500,000 live cattle from the EU to the Middle East. This trade causes immense animal suffering and yet, because of GATT's Article XI, the EU is powerless to ban it – unless it could justify such a ban under one of the exceptions set out in Article XX, discussed later.
- Articles I and III of the GATT are designed to eliminate discrimination in international trade. Article I, often referred to as the Most-Favoured-Nation clause, prohibits a country from discriminating between different foreign nations. Contracting parties must not grant any advantage, favour or privilege to a particular product from one country unless it is also granted to like products from all other countries. In short, all countries must be treated alike.
- Whereas Article I prevents a country discriminating between different nations, Article III prohibits discrimination within a country between the products of domestic producers and

imported products. In short, countries must not protect domestic production. In particular, Article III (4) provides that imported products must be given treatment no less favourable than that accorded to like products of domestic origin.

At first sight Articles I and III appear to pose no major threat. Surely a country is entitled to prohibit the marketing of, for example, battery eggs provided that this prohibition applies with equal force to both domestically produced and imported battery eggs. Sadly, things are not so simple for WTO members. There must be no discrimination between imported products and '*like products*' of domestic origin. GATT panel reports have provided an unhelpful, perverse interpretation of the key term 'like products'. For them, 'an egg is an egg' whether it is battery or free-range and 'tuna is tuna' whether or not catching it involved the killing of dolphins.

PRODUCT IS ALL

The disputes panels have decided that in determining whether an imported product is 'like' a domestically produced one – and thus entitled to equally favourable treatment – no account may be taken of *the way in which* it was produced. Processes and production methods must be ignored; only the actual end product may be looked at.

This is a major problem as most attempts to improve animal welfare are focused on changing *the way in which* an animal is reared or caught rather than on the end product. At WTO, however, no distinction may be made between battery eggs and free-range eggs. Thus a country cannot prohibit the marketing of battery eggs – if that prohibition extends to imported eggs – because, WTO decrees, a battery egg is a like product to a free-range egg and cannot be treated less favourably than a free-range egg.

These problems were highlighted in the two Tuna–Dolphin cases. In the Eastern Tropical Pacific Ocean schools of tuna tend to swim beneath dolphins. Fishermen often use the dolphins as a way of locating the tuna. Purse seine nets are cast around both schools with many dolphins being trapped and dying in the nets. In an attempt to protect dolphins the US introduced the Marine Mammal Protection Act. This prohibited the import of tuna from countries which failed to meet the dolphin protection standards laid down by US law.

Mexico challenged this measure in the case known as Tuna–Dolphin I. Later a related measure was challenged by the EU in

Tuna–Dolphin II. The dispute panels ruled that the prohibition by the US of imports of tuna from Mexico was contrary to Article XI, which prohibits import bans. The panels also suggested that the US measure was inconsistent with Article III as a country is not entitled to distinguish between tuna products according to the way in which they were caught. Under the GATT, tuna caught in a way which leads to the death of many dolphins is a 'like' product to tuna caught in a dolphin-friendly manner. And like products must be treated alike.

In Tuna–Dolphin II, the dispute panel stressed that:

> ... *Article III calls for a comparison between the treatment accorded to domestic and imported like* products, *not for a comparison* of the policies or practices *of the country of origin with those of the country of importation* [my emphasis].

For animal protection, it is absurd that a country cannot prevent the import of fish whose catching involves the wanton destruction of many dolphins. GATT is now a major obstacle to animal welfare reforms because of its policy that an imported product cannot be discriminated against because a country believes that *the way in which* it was produced is cruel. Yet what a country often wants to say is, in effect: 'we wish to ban imports of this product – or the marketing of this product in our territory – because in our view it has been produced in a morally unacceptable manner.' It is outrageous that a free trade treaty is preventing countries from acting in line with their ethical values.

The GATT's approach has widespread implications. For example, the EU Directive on welfare at slaughter provides that meat imported from third countries must have been slaughtered in an abattoir which has animal welfare standards at least as high as those set out in the EU Directive. This provision is probably inconsistent with the GATT, which sees meat as meat and the way in which the animal was slaughtered is, sadly, immaterial.

Clearly, the GATT prevents countries prohibiting the import of products coming from animals reared in cruel systems. In effect, however, it also makes it difficult for a country to ban that system even in its own territory. If, for example, the EU wanted to ban the battery cage, it could not ban the import of battery eggs from countries such as the US or Brazil. In these circumstances, the EU will be discouraged from banning the cage as European egg farmers would put huge pressure on the EU, saying that their industry would be destroyed by the import of cheap battery eggs from outside the EU.

Exceptions to the rules

Article XX contains a number of general exceptions which were designed – despite the GATT's fundamental free trade goals – to allow countries to give due weight to other important public policy objectives.

Article XX provides that:

> *Subject to the requirement that such measures are not applied in a manner which would constitute a means of arbitrary or unjustifiable discrimination between countries where the same conditions prevail, or a disguised restriction on international trade, nothing in this Agreement shall be construed to prevent the adoption or enforcement by any contracting party of measures:*
>
> *(a) necessary to protect public morals;*
>
> *(b) necessary to protect human, animal or plant life or health;*
>
> *(g) relating to the conservation of exhaustible natural resources if such measures are made effective in conjunction with restrictions on domestic production or consumption;*

At first sight, these exceptions seem very helpful for animal welfare. Most animal protection measures are issues of public morality or designed to protect the life or health of animals. The common sense meaning of the exceptions has, however, been eroded over the years by dispute panel reports eager to ensure that no other considerations are allowed to impede the onward march of free trade.

Crucially, the Tuna–Dolphin I panel stated that a country's measures to protect animal life or health or to conserve exhaustible natural resources could not be extra-territorial in their aim. A country may act to protect the life and health of animals located within its own territory; in general, however, it may not act to protect animals located outside its territorial jurisdiction. In short, a country cannot take action designed – through the use of unilateral trade measures – to extend its domestic laws and standards beyond its national boundaries. In particular, measures which try to force other countries to change their laws are not protected by Article XX (b). The Tuna–Dolphin I panel stressed that if extra-territorial measures were permitted 'each contracting party could unilaterally determine the life or health protection policies from which other contracting parties could not deviate without jeopardising their rights under the General Agreement'.

The bar on measures which have extra-territorial application means that countries are effectively prevented from taking an ethical stance by prohibiting the import – or marketing – of products the production of which has involved animal suffering. The aim of countries which wish to do this is not to force other countries to change *their* standards, but rather to be at liberty to prohibit ***within their own territory*** the marketing of products (whether domestically produced or imported) derived from practices which have an adverse effect on the well-being of animals. It is ridiculous that ethical considerations must take second place to the requirements of free trade.

Moreover, to benefit from the Article XX (b) exception, a country must show that its measure is 'necessary' to protect animal life or health. GATT panels have construed 'necessary' very narrowly. In two cases – 'the US – Section 337 of the Tariff Act of 1930' case and 'the Thai cigarette' case – the panels ruled that a measure is necessary only if no alternative measure which is consistent with – or less inconsistent with – GATT rules is available.

A new attempt to impose extra-territorial measures was made by the US sea turtle laws. These ban the import of shrimp which are not caught by methods which protect sea turtles from incidental drowning in shrimp trawling nets. This measure was considered by a WTO dispute panel as Malaysia, Thailand, India and Pakistan argued that the ban is inconsistent with the GATT. As we have seen, the GATT does not allow discrimination on the basis of production methods. For the GATT, shrimp caught in a way which drowns turtles are the same product as shrimp caught using turtle excluder devices.

The US defence was based on Article XX. It argued that its measures are necessary as all species of sea turtle are threatened by extinction and that, without the use of turtle excluder devices, other measures are not sufficient to allow sea turtles to recover from the brink of extinction. On extra-territoriality, the US argued that sea turtles are a shared global resource which does not fall exclusively within the jurisdictions of the complaining countries. Appallingly, the dispute panel rejected the US arguments and ruled that its attempts to protect sea turtles are inconsistent with the GATT. The dispute panel's ruling in the turtle–shrimp case was totally uncompromising in its refusal to allow an animal protection measure to benefit from an Article XX exception and thereby take precedence over free trade. The slight hopes that had been raised by two of the cases referred to below have largely been dashed by the turtle–shrimp decision.

There had been some signs that Article XX exceptions were being taken a little more seriously and not treated as impediments to free trade. In the US Reformulated Gasoline case the WTO appeals panel – which is called the Appellate Body – stressed that Article III.4 (one of the main free trade provisions) may not 'be given so broad a reach as effectively to emasculate Article XX (g) [the conservation of exhaustible natural resources exception to free trade] and the policies and interests it embodies'.

The Appellate Body emphasized that the relationship between GATT's main free trade Articles and the policies and interests embodied in Article XX exceptions must be worked out on 'a case-to-case basis, by careful scrutiny of the factual and legal context in a given dispute...'. In short, each case should be looked at on its own merits rather than from a standpoint that the Article XX exceptions must always be quashed in the interests of free trade.

One exception not yet tested is Article XX (a) which allows measures 'necessary to protect public morals'. I believe that countries should vigorously attempt to defend measures designed to protect animals under this 'public morality' clause.

CIWF – together with the International Fund for Animal Welfare – recently brought a case at the European Court of Justice asking for a declaration that – despite the EU's free trade rules – the UK is legally entitled to ban the export of calves to veal crates. In his Opinion, the Court's Advocate-General said that where, in a particular country, matters of animal life and health have become matters of public morality, that country could ban the export of calves to veal crates. Sadly, the Court rejected this Opinion, but I would now like to see countries trying to use the public morality argument in the WTO.

REVERSING GAINS

Unfortunately, the GATT has already led to the erosion of two key EU measures. The 1991 Leghold Trap Regulation prohibits the use of such traps for catching fur-bearing animals in the EU. It also sought to ban the import of pelts from 13 species from countries using the leghold trap. Fearing that this import ban could not survive a challenge at the WTO, the EU has failed to implement it and has instead negotiated agreements with Russia, Canada and the US which allow the continued use of leghold traps subject to some rather weak conditions. The GATT has, in effect, led to the unscrambling of one of the EU's main animal protection laws.

The EU Cosmetics Directive has similarly been largely unravelled. The Directive prohibited the marketing after 1 January 1998 of cosmetic products where the product or any of its ingredients had been tested on animals after that date. Again a GATT challenge was feared and the Directive has not been brought into force, although the European Commission has promised to come back with a much more limited proposed Directive, reworded to make it GATT-compatible.

The Commission's plan is to ban only the *testing* (not the *marketing*) of cosmetics within the EU. By scrapping the ban on the marketing of animal-tested cosmetics the EU hopes to avoid a challenge at WTO on the basis that the EU is failing to treat imported cosmetics (which may well have been tested on animals) on equal terms with domestically produced cosmetics, which would not have been tested on animals. However, to ban the testing of cosmetics on animals within the EU (rather than the marketing of cosmetics tested on animals) would emasculate the original Directive as the multinational cosmetics companies will probably do their animal testing outside the EU and then import the products for sale within the EU. It may also be that the European Commission is only using the GATT as an excuse to dismantle the Cosmetics Directive, as it is a measure it never liked.

In future, I fear the EU will decline to introduce animal protection laws on the basis that they would violate the GATT. In some cases that fear will be justified; the GATT is deeply inimical to animal welfare. In other cases the EU will, I suspect, use the excuse of a potential challenge at WTO to avoid introducing measures advocated by much of the public. Indeed, when an EU-wide ban on veal crates was being debated, one reason France gave for opposing the ban was that the EU would not be allowed under the GATT to ban the import of veal from calves kept in crates outside the EU. Interestingly, the Council of Ministers ignored this danger and agreed to ban the veal crate throughout the EU. However, in its 1997 report on the welfare of pigs, the European Commission's Scientific Veterinary Committee warned that if the EU adopted measures to improve pig welfare, it would – because of the GATT – be unable to prevent the import of pigmeat from countries where pig welfare standards are lower.

In January 1998, the WTO Appellate Body ruled on the case brought by the US and Canada challenging the legality of EU prohibitions on the imports of meat from cattle to which certain hormones have been administered as growth promoters. In August 1997, the WTO dispute panel found the EU had violated one of the WTO side agreements – the Agreement on the Application of Sanitary and Phytosanitary Measures (the SPS Agreement). This Agreement deals with certain food safety measures; in particular, those to protect human and animal life or health from risks arising from additives or contaminants in foods. The panel ruled that the EU did not have scientific justification for imposing a higher level of protection than would be achieved by measures based on the relevant international standards, in this case the Codex Alimentarius recommendations. The panel also ruled that the EU had acted inconsistently with the SPS in not basing their measures on a proper risk assessment. The EU appealed against the panel's rulings.

The Appellate Body reversed two of the panel's findings in a ruling that left both the US and the EU claiming victory. The Appellate Body made it clear that 'the right of a Member to determine its own appropriate level of sanitary protection is an important right' and that that level of protection may be higher than that implied in the international standard. At the same time, the Appellate Body stressed that the setting of such a higher level of protection must be based on a risk assessment. Moreover, under the SPS any sanitary measure must be justified by 'sufficient scientific evidence'. Under the Appellate Body's ruling, the EU can carry out a further risk assessment. If that shows that there is a proper justification for concerns about the impact on human health of meat coming from hormone treated animals, the EU would be able to retain its ban.

This Appellate Body ruling continues the trend of the Reformulated Gasoline case by recognizing that considerations other than those of free trade must be given due weight in WTO. The Appellate Body emphasized the need for a 'delicate and carefully negotiated balance' between the sometimes competing interests of promoting international trade and of protecting the life and health of human beings. The WTO now needs to recognize that the life and health of animals is a factor which should be given at least equal weight to free trade. Unfortunately, animal life and health is treated as a trivial matter by the WTO.

The hormone case ruling also helped disperse the notion that a particular ban can only be maintained if there is monolithic, incontrovertible, scientific justification for that ban. The Appellate Body recognized that sometimes there will be differing scientific opinions and that in some cases 'responsible governments may act in good faith on the basis of what, at any given time, may be a divergent [ie not a mainstream] opinion coming from qualified and respected sources'. This is very welcome as in many cases to do with animal welfare there will not be complete agreement among scientists. Often the majority will agree that a particular rearing system is cruel, but a minority may take an opposing view.

Although the beef hormones case has revolved around questions of human health, there is ample evidence that some of the hormones in question have adverse effects on animal health.

BST

Similar issues arise with BST. I hope the EU will renew its prohibition on the use of BST when the current EU moratorium expires at the end of 1999. BST can have severe effects on the health and welfare of dairy cows. It can lead to an increase in the incidence of mastitis as well as to foot and leg problems and digestive disorders.

The EU does not prohibit the import of dairy products from BST-treated cows and is unlikely to try to do so. But will the EU continue to ban the use of BST *within the EU* as such a ban could be challenged under the GATT as being tantamount to a ban on the import of BST? The Appellate Body remarks in the meat hormones case should encourage the EU to retain its moratorium on the use of BST. If a ban is challenged, the EU will have to justify it on the grounds of the threat posed to animal health by BST as the EU will presumably, in part, be relying on the SPS exception allowing measures necessary to protect animals' health.

CHANGE THE GATT/WTO

Overall, the GATT/WTO has already wrought an immense amount of damage to animal protection measures. Moreover, it makes countries fearful of introducing measures designed to improve animal welfare. It also provides a respectable excuse for those who have no wish to introduce higher welfare standards.

It is extraordinary that the EU has signed up to a treaty whose sole concern is free trade. I am not opposed to free trade but it should not be accorded a higher place in our law and our value systems than ethical concerns such as the desire to prevent the cruel treatment of animals or people. How many people, if asked to nominate the values which they believe should play a central role in our world, would give free trade a high place on the list? And yet we are subject to a treaty which insists that all our values must be subservient to trade liberalization.

The time has come to stop this nonsense. We must campaign and lobby to secure amendments to the GATT/WTO which allow other values – such as the humane treatment of people and animals – to take precedence over free trade. It will be a long hard struggle but it can be done. We must let our politicians both here in the UK and in Europe know of our concerns. Many MPs are not aware of these problems. Also, we must not allow the recognition by the EU of animals as sentient beings to be empty words. It must be translated into the detailed directives of the EU. And if animals are sentient beings, not goods or products, then they should not be subject to free trade rules.

18 Penalizing the Poor: GATT, WTO and the Developing World

Vandana Shiva

The World Trade Organization (WTO) is a global, free-trade-treaty organization which, by excluding concern for humans and animals as the basis of trade and commerce, has, in effect, outlawed human rights and animal rights from commercial activity.

Despite free trade being increasingly justified as in the interest of the people of the developing world, WTO rules make the poor in developing countries pay the most because they make them bear a disproportionate burden of the social and ecological costs of globalization. Furthermore, since the survival of poor people in developing countries is intimately connected to partnership with other species, the violation of animal rights is simultaneously a violation of human rights there. The conflict between animals and humans in global trade is artificial and contrived. As three years' experience of WTO disputes and dispute rulings indicates, the real conflict is between the free trade rights of global corporations and the rights of humans and animals.

Should protecting the sacred cow be GATT-illegal?

On 23 March 1998 at a Meeting on Trade and the Environment in Geneva, Renato Ruggiero, Director General of WTO, said, 'Nothing in WTO stands in the way of the environmental agenda.'[1]

On the same day, WTO announced the initiation of a dispute by the EU against India, which in effect forces India to kill animals. To protect animals and rural livelihoods, India has a restriction on the export of raw hides and furs. The EU argues that preventing free export of furs and hides contravenes Article XI of GATT 1994. As the EU communication states:

> *The European Communities are very concerned by this practice as it appears to nullify or impair the benefits accruing to the European Communities under GATT 1994, particularly because such practice limits the access of EC industry to competitive sourcing of raw and semi-finished materials.*[2]

According to WTO rules, the right of European industry to cheap raw materials is higher than the right to life and livelihoods embodied in India's culture, India's constitution and in India's export and import policy and in international Multilateral Environment Agreements (MEAs) such as the Convention on Biological Diversity (CBD). According to Article XI of GATT, any restriction on imports and exports is illegal even though such restrictions might be necessary for cultural, ecological and economic reasons. Cultural values, conservation values and value in local and domestic economies cannot be protected through restrictions on exports or imports.

Article XI of GATT states:

> *No prohibitions or restrictions other than duties, taxes or other charges, whether made effective through quotas, import or export licences or other measures, shall be instituted or maintained by any contracting party on the importation of any product of the territory of any other contracting party or on the exportation or sale for export of any product destined for the territory of any other contracting party.*[3]

In essence, this article implies that:

- everything is tradeable, including animals and animal products; and
- any mechanism that translates into a restriction on trade is GATT-illegal.

Hence environmental policies and policies that protect the rights of animals or humans need to be dismantled under free trade rules.

We already know what this implies for animal welfare and the welfare of the rural poor in India after the experience of meat exports from India under trade liberalization pressures.

The Impact of Trade Liberalization on Animals and Rural Livelihoods

The Indian government's new economic policy of 1991 was based on trade liberalization and it encouraged the setting up of 100 per cent export-oriented meat plants. A new livestock policy was introduced to ensure trade liberalization. It recommends government interventions to stimulate meat production even though this will totally undermine the basis of sustainable agriculture.[4]

The Ministry of Agriculture has given 100 per cent grants and tax incentives to encourage the setting up of slaughterhouses. According to a 1996 Union Ministry of Environment report, at least 32,000 illegal slaughterhouses have established themselves in the last five years, compared to only 3,600 licensed abattoirs legally established. The government affirms that this is simply an estimate, which in reality is bound to be much greater.

The total quantity of meat exports has risen more than twenty-fold from 6,196 tonnes in 1975 to 137,334 tonnes in 1995.[5] Beef and veal, buffalo meat, and total meat exports have all increased by factors of 2.15, 1.93, and 1.91 respectively from 1990 to 1995. India now exports more meat than the animals being replenished within the country. This is leading to a serious decline of ecological sustainability and economic survival of rural communities.

In the last decade, there has been a significant decline of livestock in India, particularly the indigenous breeds known for their hardiness, milk production and draught power. This decline is primarily due to the illegal slaughtering of cattle and buffalo for meat exports. Liberalization of the exports of raw hides and fur skins will cause similar trends and lead to erosion of cattle wealth and extinction of animal genetic diversity. Free global trade in animal skins will encourage non-sustainable slaughter of animals to increase availability, unlike local trade in India which is dependent on animals dying natural deaths before their skins and hides are used.

The dramatic decline in livestock population in India has reached grave proportions. Without measures to arrest this trend now, we will witness the extinction of livestock within our lifetime, and with it the foundation of sustainable agriculture will disappear. Animals provide organic manure and renewable energy. Killing them for meat exports or exports of hides destroys far more in economic terms than the economic growth generated through trade.

For example, in one slaughterhouse alone, the projected export earnings are Rs 200 million. Yet the economy destroyed is equivalent to Rs 9.1 billion through the loss of manure and animal energy that would be supplied by the animals if they were allowed to live.[6] The same logic will apply to the exports of furs and skins, since furs and skins come from animals and an export trade would increase the demand for cheap skins and furs from India. Forced global trade in these animal products will create non-sustainability and rural poverty in India.[7]

THE IMPOSITION OF NON-SUSTAINABILITY AND VIOLENCE[8]

The role of animals in tropical farming systems is not fully appreciated. Most animal husbandry models come from industrialized countries where livestock are separate from crop production and are maintained for the dairy industry. As a result, indigenous breeds maintained for animal energy and draught power or for organic inputs to maintain soil fertility have been displaced.

The promotion of increased meat production for export is leading to the erosion of India's genetic livestock diversity and depletion of its cattle wealth. Large populations of indigenous livestock species are disappearing due to the increased slaughter rate for export. Furthermore, the emphasis on commercial dairy with the eventual aim to export, is threatening the existence of India's indigenous breeds due to the cross-breeding with exotic species.

In 1996, the FAO confirmed that:

> *the diversity of domestic animal breeds is dwindling rapidly. Each variety that is lost takes with it irreplaceable genetic traits – traits that may hold the key to resisting disease or to productivity and survival under adverse conditions.*

For example, some of the declining indigenous breeds today are Pangunur, Red Kandhari, Vechur, Bhangnari, Dhenani, Lohani, Rojhan, Bengal, Chittagong Red, Napalees Hill, Kachah, Siri, Tarai, Lulu and Sinhala.[9]

Recognising the erosion of animal genetic resources, Agenda 21 states the need for conservation and utilization of animal genetic resources for agriculture:[10]

> *Some local animal breeds have unique attributes for adaptation, disease resistance and specific uses which in addition to their socio-cultural value should be preserved. These local breeds are threatened by extinction as a result of the introduction of exotic breeds and of changes in livestock production systems.*

Rebuilding animal and crop diversity is an important policy aspect of sustainable agriculture. Elsewhere, Agenda 21 calls for conservation and sustainable utilization of the existing diversity of animal breeds for future requirements.[11] But since India's ratification of GATT, and the implementation of World Bank recipes of structural adjustment, India's new economic political climate has hastened the rate of depletion of our animal wealth and the extinction of our animal diversity.

The decline in animal wealth is seriously undermining the foundation of sustainable agriculture and helping to destroy the rural economy and rural livelihoods. This will adversely affect the landless, the dalits[12] and women. Women provide nearly 90 per cent of all labour for livestock management. Of 70 million households which depend on livestock for their livelihoods, two thirds are small and marginal farmers and landless labourers. Cattle exports are pushing livestock prices up, which adversely affects the ability of the small farmer community to buy them. Reduced amounts of dung for manure, cooking and fuelling biogas plants further reinforce the trend towards unsustainable agricultural systems and rural economies. Consequently, farmers become increasingly dependent on imported non-renewable fossil fuels for fertilizers and energy.

Proponents justify meat exports as a means of earning foreign exchange. Yet the destruction of the cattle wealth of the country is actually leading to economic destruction and a drain of foreign exchange through increased imports of fertilizers, fossil fuel, tractors and trucks to replace the energy and fertility that cattle give freely to the rural economy.

Through the dispute on raw hides and fur skins, the EU is using the WTO to force India to remove restrictions on exports of furs and skins. This, in turn, implies restrictions on animal protection, since furs and skins are on animals before they enter international trade as commodities. India has these restrictions in place through the export-import (EXIM) policy for both environmental and economic reasons. Environmentally, exports are restricted to protect our cattle wealth, which is the basis of sustainable agriculture, and the survival of small farmers and small craftspeople. The restriction on exports of skins and furs is necessary because of the livelihoods provided to craftspeople, shoemakers, cobblers, farmers and other small producers. Exports would divert skins to the European market, thus depriving craftspeople of access to raw material. In 1993, when India was forced to remove export restrictions on cotton, two million weavers lost their livelihoods.

In addition to the destruction of the livelihoods of craftspeople dependent on leather as raw material, free trade in animal skins would also destroy livelihoods and the economic base of agricultural communities. Whether for meat or skin exports, trade liberalization is basically a forced imposition of a violent and destructive economy on non-violent living and sustainable economies. The living, non-violent economy thrives when it is protected – through cultural values and restrictions on commerce. When such restrictions are removed, the living economy has to compete with the death economy. Since the violent and destructive economy is backed by more powerful interests and generates higher profits, it destroys the non-violent living economy.

For example, 100 per cent export-oriented slaughter houses get the following subsidies for:

- transport – 25 per cent;
- pre-cooling – 50 per cent;
- water treatment – 50 per cent;
- cold storage – 100 per cent;
- product literature – 40 per cent;
- brand publicity – 40 per cent;
- packaging – 60 per cent;
- package development – 30 per cent;
- quality control – 50 per cent; and
- assistance for air freight – 25 per cent.

I work with organic farmers in India; we cannot get enough loans or subsidies to buy a pair of bullocks so that the farm can run organically – the economy based on killing animals gets all the subsidies. The indigenous culture and economy that has managed to survive over centuries, that has ensured that the soil survives, that the animals survive, that the human communities survive, has now been dismantled. Exports of animals and meat have shot up by 2000 per cent as a result of trade liberalization. It is having an absolutely devastating effect on soil fertility management and is destroying the livelihood of people who depend on cattle in the local economy. Rawhides and furs are on animals before they reach the market. This pressure to export basically is a pressure to distort the local economy so that it pays more to kill and it punishes you to keep your cattle alive. But WTO affects life offshore too.

Killing turtles and traditional fishing communities

The WTO ruling in the shrimp-turtle dispute – over how shrimp are caught – has protected the worldwide diffusion of non-sustainable industrial fishing technologies such as trawlers to meet the non-sustainable luxury demands of rich consumers in the North. It has declared the protection of species and people's livelihoods as illegal from the perspective of free trade. WTO has ruled against the struggles of environmentalists and fishworkers in India who are trying to protect marine and coastal ecosystems, not merely against the US environmentalists. WTO has put profits and luxury consumption above the right to life of turtles and the right to survival of traditional fishing communities in India.

WTO has institutionalized through its rules and its rulings an agenda that puts profits above life. This agenda must be reformed if life and living resources are to be protected. In this new era of defending the environment under globalization, a new solidarity and cooperation is needed between environmental movements in the South and in the North. This new solidarity should also recognize that the real conflict is not between the people and the turtle but between the trawler and the turtle, and global trade and the turtle.

Protection of the turtle, therefore, needs to be linked to protection of traditional fishing communities and their culture of conservation by strengthening environmental laws which protect both the environment and people. Dismantling environmental regulation can only accelerate environmental destruction. Since environmental deregulation is an essential part of trade liberalization, free trade and the protection of the environment cannot coexist. If the turtle has to be saved, destructive trade and the use of destructive technologies need to be stopped.

Violent and Wasteful Technologies of Large Scale Commercial Fishing

In India, the turtle is considered sacred. It is one of the ten '*avatars*' or incarnations of Vishnu, the lord of creation and maintainer. The Satapatha Brahmana states, 'The Lord of progeny, having assumed the form of a tortoise, created offspring. He made the whole creation, hence the name Kurma given to the tortoise.'[13]

In the myth of *sagar manthan* or the churning of the oceans, the God Vishnu appeared in the form of the turtle to recover some of the things of value lost in the deluge in the earlier '*yuga*' or era. The churning could take place only when Vishnu as the turtle went to the bottom of the ocean. The turtle became the pivot on which Mount Mandara rested as the churning stick. This symbolizes the significance of the turtlc in sustaining life. That is why villagers along India's coasts relate to turtles with respectful reverence. Traditional fishing communities use non-violent technologies to ensure that marine species like turtles are not killed or hurt. People and turtles have coexisted along India's coast over centuries.

This coexistence has been threatened by the introduction of violent technologies such as trawlers. An estimated 150,000 turtles drown each year when they are caught in the nets of large trawlers. Such mechanized trawlers have been introduced in the Indian waters over the past few decades in the name of 'modernization' of fisheries through development financing.

More specifically, in South Asia, bottom trawling, which was introduced in a big way in thc 1906s, helps primarily to increase the production of shrimp which are exported to Japan and the US. Shrimps are generally found in shallower inshore waters. Using this capital-intensive technology to fish for shrimp comes into direct conflict with harvesting fish inhabiting the same ecosystem which go to flavour the rice of the rural masses of the region.

Until the end of the 1950s in the South Asian region, the marine fish harvest increased at a rate of 5 per cent per annum in spite of the lack of new harvesting technologies. During this period, between 5000–6000 tonnes of prawns from India were exported to Burma, Thailand and Malaya every year in dry form and accounted for 25 to 30 per cent of the annual export value of around US$11 million (1958–59 average).

By 1976–83, after three decades of planned fisheries development in the region, the rate of growth of marine fish harvest had dropped to

2 per cent per annum. It was also during this period that the conflicts at sea were most rampant. However, during this period of relative stagnation, the exports of prawns – all destined for the Japanese and American markets in frozen form – increased dramatically.

Excessive bottom trawling of inshore waters, something which is inevitable in the pursuit of shrimp, is tantamount to a continuous raking of the seabed, causing murky and turbid waters. This destroys the abodes of young demersal fish and bottom dwelling spawners.[14]

Technologies such as purse-seining and trawling are powerful but destructive. Highly capitalized trawler fleets and purse-seiners use nets which scoop up whole shoals of fish, many of which are not of commercial value but have high ecological value. Trawlers do not merely kill turtles, they are also a threat to marine biodiversity. Species which do not have commercial value on global markets, or are of the wrong size for standardized marketing and packaging, are just killed and thrown back into the sea. These are called 'by-catch' and 'discards'. As the *Ecologist* reports, annual global discards in commercial fisheries have been conservatively estimated at 27 million tonnes, equivalent to more than one third the weight of all reported marine landings in commercial fisheries worldwide.[15]

A study from Alaska suggests that Bering Sea red king crab discards amounted to more than five times the number actually landed. In cod fishery in Norway, the waste over one season in 1986–87 was 100,000 tonnes. In 1986–87, two billion kilograms of fin fish were dumped overboard. Worldwide, the shrimp and prawn trawler fisheries are reported to have the highest level of 'discards' of any fishery – about 16 million tonnes a year. In some shrimp fisheries, up to 15 tonnes of fish are dumped for every tonne of shrimp landed. Most of this by-catch is thrown back either dead or dying. Turtles are among those 'unwanted species'.

The species diversity which is treated as waste by global commercial fishing fleets is the economic base for the traditional fisherman. It is the ecological base for maintaining the webs of life and food chains that create the sustainability of the marine environment.

When productivity is assessed in the wider context of diversity of livelihoods, species diversity and the future, the 'efficient' technologies of industrial fisheries chasing the reductionist value of maximizing commercial catch in the short run are clearly rather inefficient. Over-capitalized fisheries are creating fisheries collapse in region after region. Nine of the world's major fishing grounds are threatened. Four have been 'fished out' commercially. Total catches in the North West Atlantic have fallen by a third over the past 20 years. In Newfoundland, fishing grounds were closed indefinitely in 1992. In Europe's North Sea, the Bering fishery had to be closed from 1970–1972. According to the FAO, which in 1991 had said that global fish catches would continue to increase, now an estimated 70 per cent of global fish stocks are 'depleted' or 'almost depleted' and 'the oceans' most valuable commercial species are fished to capacity'.

The misplaced efficiency of technologies created in response to maximizing commercial catch has the social impact of destroying the livelihoods of traditional fish communities through the ecological impact of undermining the very basis of sustaining fisheries activities.

Indian calls for ban on trawling and industrial aquaculture

Because of the intrinsically destructive nature of the technology, the traditional fishing communities in India have been calling for a ban on mechanized trawlers since the 1970s, both to protect marine life and their livelihoods. They have also succeeded in preventing new licences being given for deep sea fishing vessels. Coastal communities, including fishing communities, have also resisted industrial shrimp farming which is devastating the coastal zone.

We have fought against the shrimp lobby which has destroyed India's coastal ecosystems through the cancerous spread of industrial shrimp aquaculture. In 1996, the Supreme Court ordered a ban of industrial shrimp farms within the Coastal Regulation Zone (CRZ). However, the shrimp export lobby has spared no effort in trying to undo the Supreme Court Judgement as well as the National Environmental Regulation for protecting the fragile coastal ecosystems. Such environmental deregulation is an inevitable consequence of trade liberalization and commerce-led policy.

Some of the shrimp harvested or farmed in India is exported to rich countries. The best step for Northern consumers to ensure sustainable consumption would be to act in solidarity with movements in India calling for a ban on industrial fishing and industrial aquaculture and to boycott the consumption of shrimp harvested by mechanized trawlers or farmed through non-sustainable aquaculture. This would, of course, involve a reduction in consumption by the rich and a reduction in global trade, but it would rejuvenate marine resources and the livelihoods of fishing communities.

US calls for TEDs and trade actions

Instead of taking this step based on solidarity and genuine sustainability, the environmental community in the US focused exclusively on Turtle Exclusion Devices (TEDs), rather than a ban on trawling, and government action instead of consumer boycotts. As a brief prepared by US groups in the shrimp–turtle dispute acknowledges:

> *The US is one of the two largest consumers of shrimp products in the world and its shrimp consumption is a major cause of turtle deaths.*
>
> *Given the causal connection between shrimping and turtle mortality, the US ability to reduce the impact of its shrimp consumption on sea turtles is critical to protecting endangered sea turtle populations. The use of TEDs in shrimp trawls that serve the large US market represents the most environmentally sound and effective method available to the US to protect these endangered species while allowing human shrimping activity to continue relatively unimpeded.*
>
> *TEDs are essential to the adequate protection of sea turtles.*[16]

While the US groups wanted to allow 'human shrimping activity to continue relatively unimpeded', the Indian movements recognized that industrial overfishing was itself non-sustainable and needed to be reduced and stopped. This would also reduce luxury consumption of shrimp, a sacrifice US consumers did not seem to be ready to make. Therefore, in the 1990s, when the US environmental community became concerned about the unnecessary killing of turtles by trawlers, they called for the use of TEDs so that the turtles could escape if caught. US environmental groups pressured the US government to ban shrimp imports if the shrimp was caught by vessels not using TEDs.

While this step was well-meaning, it was ecologically inadequate and politically inappropriate. As a means of protecting marine biodiversity, the US ban is neither necessary nor sufficient. The ban on trawling as demanded by Indian movements would be far more effective than the trade conditionality linked to introducing TEDs through a ban on shrimp imports. Since all shrimp caught is exported to the

US, Europe and Japan, if Northern consumers do not want to consume shrimp that is caught in environmentally destructive ways, they should stop consuming all shrimp caught through industrial fishing technologies such as trawlers and produced through industrial aquaculture. Such a ban would not reflect the politics of unilateralism; it would be an expression of solidarity since such a ban on trawling has been called for by the fishing communities and environmental community of India.

From the perspective of the Indian environmental and fisheries movements, the US groups did not go far enough in the direction of sustainability and democracy. By using trade sanctions linked to TEDs through the US government, US environment groups have used an environmentally weak and politically undemocratic means to attempt to save the turtle. However, the inadequacy of means used by the US environmental community does not justify the WTO's role in protecting the forces that pose a threat to the survival of turtles and other species, including humans.

WTO has ruled against the US in the shrimp import dispute. Asian countries, including India, Malaysia, Thailand and Pakistan, initiated a dispute against the US on grounds that it was acting unfairly by banning imports of shrimps caught in nets without TEDs.

The shrimp ban was not the most effective step to save the turtle. It also exhibited a total lack of awareness that, through their strong movements, traditional fishing communities had gone much further than US environmentalists in calling for a ban on trawlers, not merely the introduction of TEDs in trawler nets. However, WTO's ruling is indifferent to the environmental aspects of the ban. It has merely focused on the trade dimensions. Since all environmental regulations restrict environmentally destructive commerce, according to the WTO they are trade restrictive, hence GATT-illegal. The WTO ruling in the shrimp–turtle case encourages the killing of turtles and goes against the movements of citizens of both India and the US.

The threat posed to turtles by mechanized trawlers is a major one: 1998 was the second year in a row that turtles did not come for mass nesting to Gahirmata beach in Orissa, along the Bay of Bengal.[17]

WTO's ruling is a victory of trading interests who have no loyalty to any country or any ecosystem. Its decision is not a victory for the movement of traditional fish workers and the environment movement in India, which is fighting the marine export lobby and its use of non-sustainable technologies not just at sea but also on the coast. It is not a victory for India because India is not the global shrimp industry, which does include exporters from India. India is her coasts and marine life, her mountains and rivers, her farms and forests. India is the peasants and tribals and fishworkers whose resources and livelihoods are being destroyed by destruction of the environment. India is her turtles.

The turtle dispute reflects a conflict between the environment and trade, between the turtle and the trawl, between traditional fishing technologies and violent 'modern' ones, between the 'right to life' of fishing communities and turtles on the one hand and the 'right to trade' of commercial interests who are driven by profits alone on the other. It is an example of a bigger concern.

WTO AGAINST THE RIGHT TO LIFE OF ALL SPECIES

Trading interests and trade bureaucrats in WTO repeatedly state that GATT is a 'rule-based system, not a power-based system'.[18] However, as the dispute initiated by the EU against India shows, WTO is a power-based system which puts the raw material needs of powerful European industry above the right to life of animals and small producers in India. Protection of animals is crucial to the sustainability of agriculture as well as our wildlife. Increased trade in hides and furs would mean increased slaughter of both domesticated and wild animals.

In its present form, WTO forces countries to protect the rights of powerful international interests above those of the right to life of humans and other species. This rule of profits destroys the rule of compassion and justice. India has encouraged the rule of compassion through her culture by making her animals sacred. The Indian Constitution makes compassion for all living creatures a duty of citizens and the state. The Constitution also ensures the rule of justice by enshrining the right to life and livelihood as a fundamental right of all citizens under Article 21 of the Constitution. India is a member of the international treaties such as the Convention on International Trade in Endangered Species of Wild Fauna and Flora (CITES) and the CBD which oblige countries to protect their animal diversity both domesticated and wild. These ruled-based systems, which protect our cultural values and our animals and plants, are the basis of our ecological and economic survival. But they are being declared illegal under the new rules of free trade as institutionalized by GATT and WTO.

GATT's Article XI undermines India's national Constitution which makes it a fundamental duty of the state and citizens to have compassion for all living creatures. It also violates Article XXI of the Indian Constitution which makes the protection of life and livelihoods of citizens a fundamental duty of the state. Article XI also violates two international treaties, CITES and the CBD, which oblige us to protect our animal biodiversity, since exporting hides and furs would force the killing of animals beyond the limits of sustainability, would erode our animal wealth and push species to extinction.

GATT has established an economy of violence against humans and other species caused by privileging an economy of greed over an economy of care. GATT establishes a rule of commerce and destroys the culture of compassion. The GATT rules violate the ethical rules of compassion towards other species. As the Iso Upanishad has stated:

> *The Universe is the creation of the Supreme Power meant for the benefits of (all) creation. Each individual life form must, therefore, learn to enjoy its benefits by farming a part of the system in close relation with other species. Let not any one species encroach upon others' rights.*

GATT rules, in contrast, state that powerful commercial interests have absolute rights to extinguish the rights of people and the rights of other species if these fundamental rights restrict commerce and profits in any way. Either WTO will force all countries to dismantle their systems of protection of people and animals by globalizing greed or people will force WTO to dismantle its protection of commercial profits by globalizing compassion. If the right to life of all species has to be protected, compassion has to find a place in human affairs. Article XI must be reformed to reflect values of compassion and the right to life.

A GATT FIT FOR LIFE

In the present version of Article XI, profit is the only value, trading without limits is the only right. To conserve nature and culture is GATT-illegal. To respect the higher moral order based on compassion to all beings is GATT-illegal. To respect our national constitutions is GATT-illegal. In other words, life itself is illegal. GATT is, therefore, an anti-life trade treaty and must be changed if the right to life of humans and other species has to be guaranteed.

The rule of deregulated commerce is a rule of cruelty, as every decision of the WTO reveals. Free trade rules mean that unless and until you can adopt the most violent, most non-sustainable, most unjust systems of production and consumption, you are GATT-illegal.

This rule of profits destroys the rule of compassion and justice. We in India have, over the centuries, encouraged the rule of compassion through our culture, through our laws, including our export and import laws which are being ruled as illegal under the new dispute. We now have an absolute head-on clash between rule-based systems that protect life and defend the right of life of humans and all species on earth and a rule-based system that protects profits at any cost. This contest between these two sets of rules cannot be won if we allow compassion to be turned into a footnote that will depend on interpretation on the basis of three trade lawyers.

I analysed membership of the panel that ruled against India on Intellectual Property Rights and it is unbelievable. If governments ran that way, it would be called a crony government. Now it looks as though we may have a world government run on the basis of cronyism where everyone appointed into WTO appears to have links with or sympathies for big business. As the Monsanto person said when they admitted they drafted the TRIPs agreement, 'We were the patients, the diagnostician and physician all in one.'[19]

When that kind of convergence of power starts to happen we are in a highly unstable social and political condition. We, therefore, must reinvent the rules of commerce so that they are subjected to the rule of compassion. We do not have to invent them from scratch, we just have to remember them and give them our support – to remember the Iso Upanishad teaching and not let any one species encroach upon others' rights. It is time to put the rights of animals and the rights of humans at the heart of trade treaties. That means Article I, Article III, Article XI of GATT/WTO need to be rewritten with human rights and animal rights language, not just as footnotes of what will be counted as 'sound science' in panels set up by the industry. If we fall into the trap of lobbying on the basis of footnotes and additional clauses, we have already lost.

We must address free trade from ethical fundamentals and say you made an error in Article I, you made an error in Article III, you made an error in Article XI, because this is not the kind of non-discrimination that makes life liveable. We have to learn to discriminate once more between the small and the big, the good and the evil, cruelty and compassion. Together, in this amazing moment of history, we could actually replace the rule of commerce enshrined in GATT, with new rules in which solidarity emerges between the rich and poor of the world and all species on this planet.

NOTES

1 Renato Ruggiero's speech at 'Policing the World Economy' Conference held at Geneva 23–25 March 1998
2 WT/DS120/1, World Trade Organization, Geneva, 23 March 1998
3 WTO (1994) *The results of the Uruguay Round of Multilateral Trade Negotiations – The Legal Texts*, Geneva, p 500
4 Section 3.10, New Livestock Policy, Government of India, 1995
5 The Hindu Survey of Indian Agriculture 1996, p 115
6 Supreme Court case 2267/90 on Alkabeer as well as 'No to Slaughter House' by Viniyog Parivar Trust, 1995
7 Maneka Gandhi, 'The Crimes of Alkabeer', People for Animals Newsletter, May 1995
8 Vandana Shiva, Afsar H Jafri, Gitanjali Bedi, (1997) 'Ecological Cost of Economic Globalisation: The Indian Experience', Research Foundation for Science, Technology and Ecology
9 FAO, 1996
10 In Chapter 6, para 76 of Agenda 21
11 Chapter 14.65 in Agenda 21
12 Dalit is the term used by the 'lower' castes for themselves. In official jargon they are called 'scheduled castes' because of the British schedules for population figures and census records
13 The first mention of this incarnation is found in the Satapatha Brahmana; it is also mentioned in the Mahabaratha (1.18), the Ramayana (1.45), and the Puranas (Agni Purana, ch 3; Kurma Purana, ch 259; Vishnu Purana 1.9; Padma Purana 6.259; as well as in the Bhagavata Purana)
14 Vandana Shiva (1991) *Ecology and the Politics of Survival Conflicts over Natural Resources in India*, Sage Publications
15 *Ecologist Asia*, Vol 3, No 4 July/August 1995
16 Brief prepared by US groups
17 Note on Turtles by Banka Behary Das, April 1998
18 Renato Ruggiero's Speech at 'Policing the World Economy' Conference held at Geneva 23–25 March 1998
19 James Enyard, 'A GATT Intellectual Property Code', *les Nouvelles*, June 1990 pp 54–56

19 Sustainable Agriculture's Friend and Foe: the WTO

Hugh Raven

As the WTO develops as the enforcer of the GATT trade liberalization regime agreed in 1994, there are legitimate fears that it could be or is inimical to many standards which we hold dear. Indeed, many aspects of trade liberalization can damage sustainability. But liberalization may also be beneficial – as, for example, when it attacks perverse subsidies.

WTO AS BENEFACTOR

One example of how subsidies damage sustainability is the world's commercial fishing fleet. The subsidy it receives is worth almost the same as the value of its catch. That means the fleet makes a tiny net contribution to the world economy, and does so at huge cost to marine ecology. Energy subsidies are another case – both with nuclear power, which has been an economic disaster, and fossil fuels like coal, adding to the problems of climate change and acid rain. To use economic instruments to encourage such environmental damage is madness – and if the WTO can oblige us to come to our senses, then it is a force for good.

The EU's CAP is another example, as organic agriculture shows. Organic agriculture is the closest proxy to sustainable agriculture practised on any scale in Europe, and is also the system with the highest guaranteed animal welfare standards. The UK market for organic products is very healthy at the moment, and growing fast, yet only about 30 per cent of the organic food consumed in the UK is produced here. Much of what we import could be grown in the UK, and, to encourage that, the Organic Aid Scheme pays farmers an average of £50 per hectare per year to convert from conventional farming on the first 300 hectares per farm.

Yet less than 1 per cent of UK farmland is organic. There are various reasons for this, but the most important is that organic farming cannot compete with the existing subsidies cascading into conventional farming, as two examples show:

- A well-known conventional arable farmer in East Anglia admits that he gets a cheque direct from Brussels each year for about £200,000 for his 2000 acres of arable land. The payments are under the Arable Area Payments Scheme, to compensate him for the price cuts following the CAP reforms in 1992, and for set-aside. Only the land growing eligible crops qualifies, but in this case that covers the whole farm, as it is a specialist arable unit.

  To convert to organic farming, at least half of the farm would have to be put down to fertility-building crops, such as grass or legumes, which fix nitrogen. Since they do not qualify for the subsidy under the Arable Area Payments Scheme, that would cost the farmer £100,000 in lost income each year. To compensate him, he would receive the conversion payment of £50 per hectare, but only on the first 300 hectares – a total of £15,000. The net loss of subsidy of £85,000 is a pretty big disincentive to going organic. That farmer has said he will not be converting.
- On a completely different type of farm, an extensive upland livestock unit, an average of two thirds of a farmer's income comes from livestock subsidies – Hill Livestock Compensatory Allowances, Sheep Annual Premium, Beef Special Premium and Suckler Cow Premium – all of which are headage payments. Farm income is therefore fixed directly to stocking levels.

  Because the prophylactic use of veterinary medicines has no place in a sustainable farming system – and is banned by organic standards – conversion to organic husbandry would require significant reductions in stocking rates, to reduce the incidence of parasitic worms. On the farm in question, this would mean a cut in farm income of about half. The average level of payment available under the Organic Aid Scheme in upland areas is £11 per hectare, up to a maximum of 300 hectares – or, in this case, about one thirtieth of what would have been lost through conversion.

Whether or not you agree that organic farming is a proxy for sustainable agriculture, the current pattern of very generous farm subsidies clearly locks farmers into existing production patterns. Farmers may want to convert to more sustainable systems – as they do in both these real cases. But if they did so they would incur enormous financial costs, so they do not do it.

Both these types of subsidy are directly linked to production. The Arable Area Payments Scheme is a hectarage payment, based on the crops grown in a base year. The more hectares the farmer had planted to eligible crops, the higher the payment – provided only that he goes on growing the eligible crop. In other words, the size of each year's payment, up to a given level, is directly related to the number of hectares planted – a direct link to production. Similarly with the livestock farm. The size of the payment depends entirely and exclusively on the size of the flock or herd of breeding livestock. Since breeding livestock breed, again the subsidy is related to output.

In this case, the WTO can help. When subsidies encourage unsustainable behaviour, trade liberalization can supply the remedy. The three objectives of the 1994 WTO Agreement on Agriculture are:

- to reduce production-related subsidies;
- to reduce export subsidies; and
- to open national markets to imports from overseas.

The EU managed to get a complicated agreement in the Uruguay Round GATT negotiations exempting the above subsidies from reductions. But the so-called 'peace clause', exempting them from challenge, expires in 2003. The EU is already on notice from members of the Cairns Group – a collection of the largest agricultural exporting countries outside the US and EU – that they intend to use the next round of WTO negotiations to liberalize further agricultural trade.

The Cairns Group is usually led by Australia, which recently submitted a memorandum to the UK House of Commons Agricultural Select Committee, saying:

> *members of the Cairns Group have the overall objective of placing agricultural trade on the same basis as other areas of world trade, so that agriculture is governed by the generally-applicable rules of the WTO with no exemptions or exceptions. Australia will be looking specifically for a move away from production-linked support.*[1]

If that pressure for liberalization can help to sweep away the ridiculous perversities of the CAP, then it will do us all a favour – consumer, animal welfarist, taxpayer and environmentalist alike. It would remove the biggest single impediment to growth in sustainable farming in northern Europe.

Damaging Liberalization

If it is surprising that WTO could have such positive effects on sustainable agriculture, it is because the CAP is an exceptionally silly policy. Liberalization also brings with it threats to sustainability. Other contributors illustrate how the GATT undermines animal welfare and removes autonomy from developing countries. (See Chapters 17, 18, Stevenson, Shiva.) To these I must add three areas where WTO threatens sustainability: risks to incentives to sustainability; risks to taxes or charges aimed at internalizing costs of production; and risks to consumer protection.

Incentives to Sustainability

Around the world, policies are slowly being put in place to move agriculture onto a more sustainable footing. In a very modest way, the EU is leading the way. Under the agri-environment programme, member states are introducing schemes like organic conversion subsidies. In America, the Conservation Reserve Programme pays farmers to remove land from row crops and put it under grass. In Australia, there is the Landcare programme; and so on.

Potentially, all of these schemes could fall victim to trade liberalization. Such incentive payments are vulnerable to challenge in the WTO disputes procedure as production-related support. They may not directly encourage greater output, in the way that price supports or livestock headage payments undoubtedly do, but it is difficult indeed to design a scheme which has the effect of making farming more economic which does not also have an impact on output. In the final analysis, if the farmer is still farming, he is presumably producing agricultural products. All such schemes, under that tight definition, are production related.

Although theoretically these schemes could be vulnerable, the Agreement on Agriculture contains text permitting expenditure on environmental objectives – but there is great uncertainty about how that would be interpreted. At present, the existing levels of production-related support are so high that any of the environmental schemes currently in operation are insignificant and incidental in comparison. Indeed, the definition of production-related support used in the Uruguay Round excludes the schemes mentioned above, as there are so many more obvious culprits. The likelihood of them becoming a target for trade liberalizers in the near future seems remote – but as the more obviously trade-distorting subsidies are reduced, in time it could happen.

Cost internalization

Under a different set of policies, several countries – such as Norway, Sweden, Austria and Denmark – and some US states, are attempting to internalize the costs of conventional agriculture by taxing some agrichemical inputs. These are no concern of the WTO. However, they could be vulnerable to different pressures arising from trade liberalization, in that any tax levy adds to farmers' costs, and therefore potentially undermines the competitiveness of the country's agriculture on world markets.

Concerns about competitiveness are a major brake on increasing animal welfare standards through regulation. The same argument applies to such input charges, though these are market mechanisms rather than regulations. Any discussion of eco-taxes being applied to UK farming is shouted down by the farming unions on the grounds that it would be grossly unfair, in a single market, to apply them unilaterally. This conveniently ignores their existence in other countries within that single market, and that revenues could be recycled back into the industry, so making them revenue neutral. But so far it has achieved its objective, and policymakers have rejected eco-taxes because of concerns about competition.

Consumer protection

Consumer protection is already threatened by WTO. The GATT's provisions on Technical Barriers to Trade (TBT) only accept trade restrictions through technical regulations if they seek to protect human, animal and plant health or the environment, and aim to ensure that national rules do not constitute unnecessary barriers to trade.[2]

Unfortunately, the list of permissible reasons to restrict trade does not include the provision of information to consumers as a 'legitimate objective' for regulation. As yet there have been no challenges to trade restrictions under this provision, but it is having an impact.

Many people regard GE crops as a mortal enemy of sustainable agriculture. Briefly, there is a strong argument that increased agrichemical inputs, and the spread of resistance from crop plants into other species, pose very severe risks. The EU is considering a proposal to require labelling of all genetically modified foods, but the fact it could be challenged under the GATT is being used as an argument against it proceeding through the EU policy-making process. In other words, the risk that there could be a successful challenge in the WTO is already acting as a brake on improvements in consumer standards.

The same fears apply to any compulsory labelling schemes about the method of production. If the EU wished to require products to carry, for example, information about pesticide or pharmaceutical use, it would be subject to a challenge if, in fairness, it required that of imports as well as domestic products – and it might well be prevented from having them.

The risk of a hostile ruling in the WTO is already a huge latent threat to moves towards sustainability. Yet we are entering an era when national governments increasingly recognize the limitations of their power – and as a consequence consumer-driven change is becoming ever more important. The growth of organic food, improvements in forest management through the Forests Stewardship Council, certification of fish through the Marine Stewardship Council, and the developing markets in renewable energy through harnessing consumer preference will be some of the most significant innovations in sustainable development over the next few years.

So long as they are voluntary schemes, they are safe. But any requirement to label according to production methods could well fall foul of the TBT regulations. It is in this threat to improved consumer information that the GATT poses its biggest challenge to sustainability.

CONCLUSION

In the short term, trade liberalization in the EU will be good for sustainable agriculture, as it must remove some of the perverse and ridiculous incentives offered to farmers. Beyond those short-term benefits, however, further liberalization could be damaging. If the definition of trade-distorting subsidies is tightened, many of the incentives towards sustainable agriculture could be challenged. Taxes and charges to internalize government costs will be vulnerable to concerns about competitiveness on world markets.

The biggest threat to sustainable agriculture probably lies in untested aspects of the TBT provisions. Consumer information through labelling has huge potential to shift consumer patterns in favour of sustainability. For most of us – all but the very poor – our food is freedom, a weekly, daily or hourly chance to express our 21st century destiny as consumers. Food – fast or whole, grazed or feasted – is one of our most frequent expressions of choice. In the absence of definitive case law, the latent threat of the WTO challenge will reduce the amount we are entitled to know about what we eat. I believe it is in limiting our chance to show our preferences in the market-place that GATT poses the greatest threat to sustainable agriculture.

Notes

1 House of Commons Select Committee on Agriculture, session 1997–8, Second Report, February 1998, Appendix 48, Memorandum submitted by the Australian Government

2 See 'Farm Policies and Our Food: The Need for Change', National Consumer Council, London, 1998

20 Multilateral Investment: an Agreement Too Far

Barry Coates

We have come to count the price of everything and the value of nothing.

Oscar Wilde

Powerful movements in the world are shaping our societies. And they are in conflict – not across the traditional political divides of North/South, environment/development, human rights/animal welfare, but across the divide of values. There is a fundamental conflict between the forces propelling us towards an unregulated, liberalized market economy and the countervailing forces that are aiming to strengthen development, social and environmental rights.

The building blocks for the liberalized market are being put into place extremely quickly. Two of the supporting pillars are well known, but the third is still in the shadows:

1 The international financial institutions, notably the International Monetary Fund (IMF) and the World Bank, are one pillar of the liberalization 'project'. These institutions were established in the post-World War II period as the so-called 'Bretton Woods' institutions, named after the town in New Hampshire, US, where the agreement was signed. The IMF's main role was to regulate the international financial system, specifically to make sure that the kind of financial instability that preceded the Great Depression would not occur again. Subsequently, the IMF has reinterpreted its regulatory role and has acted to deregulate capital flows. The

rapid expansion of speculative investment internationally has resulted in a series of mini-crises, mainly affecting developing countries. We now face the potential of a major financial crisis. The untold story is not the prospect of a recession in the UK or other rich countries, but the wrenching changes that have destroyed the jobs and lives of people in developing countries.

Nowadays around £1 trillion per day is transferred at a keystroke by traders in the financial markets. Capital controls around the world have had a long history, but their abolition is recent. Italy and France only abolished controls in 1990. Meanwhile Chile still has controls, widely lauded by economists. But the IMF is dedicated to removing all such controls, even in the face of evidence that speculative capital is destabilizing economies around the world. When there are problems, as shown in Mexico and Asia, the banks are bailed out while the burden inevitably falls on the poor and vulnerable. More than 100 million people in Asia alone have been cast into absolute poverty as a result of the crisis.

Meanwhile, the World Bank has joined with the IMF to liberalize national policies. The introduction of Structural Adjustment Programmes (SAPs) simultaneously in virtually every one of the poorest countries has been in response to the debts incurred during the 1970s and 1980s, when willing lenders were eager to earn high interest from loans to poor countries. When interest rates rose and oil prices increased, the IMF and World Bank stepped in with harsh economic conditionality. This ensured the banks and governments from the wealthy countries would continue to receive repayments, but placed the burden of adjustment on the poor and on the environment. When SAPs required that countries pursue export-oriented growth, the supply of a small number of primary commodities increased. However, without a corresponding increase in world demand for those commodities, the prices fell even further. The result? Africa's GNP fell by 20 per cent in the 1980s as commodity prices fell by 50 per cent. Together with social hardship caused by restructuring of government taxation and expenditure, SAPs have affected the lives of many millions of poor and vulnerable people. Poverty, suffering, ill-health and even lower life expectancy have resulted in some African countries. As with the IMF's capital liberalization policies, SAPs have treated social and environmental issues as the unfortunate side effects of economic policies designed to give primacy to tight money supply, low inflation and balanced budgets.

2 The WTO provides the second pillar of international liberalization. Together with the rules agreed under the GATT, the WTO has set itself a clear agenda, to liberalize world trade. However, a free trade system is rarely a fair trade system. As a result, social, environmental, health, animal welfare and other considerations are systematically excluded and the interests of developing countries are ignored.

The current trading system serves the interests of a small minority. Industrialized systems of agriculture and protectionism in the North mean that subsidies amount to around £11,000 per farmer each year, as much as many farmers in the poorest countries would earn in a lifetime. But under the latest agreement, the GATT Uruguay Round, African farmers would lose even more, estimated at $1.2 billion per year. The poorest developing countries still face tariffs 30 per cent ***higher*** than industrialized countries.

The rules of the WTO ensure that economic liberalization is accorded primacy over social and environmental considerations. WTO rulings on issues such as cattle hormones, bananas and shrimp–turtles clearly demonstrate the way in which non-economic issues can be overruled and how the decisions of national governments can be challenged. In the first year of WTO, some 130 cases were brought before its dispute settlement panel. Further, the related WTO agreements are extending trade law into new areas such as extending corporate control over intellectual property, thereby allowing the rip-off of indigenous peoples' knowledge and their biological resources.

3 The third and final pillar of liberalization is being put into place very quickly and quietly. It is the Multilateral Agreement on Investment (MAI). The MAI is being negotiated in the wealthy countries of the OECD, and was planned to be signed at the end of April 1998. A powerful and growing grassroots campaign around the world ensured that OECD governments could not complete the agreement, but the pressure is still on to get an agreement by May 1999.

The MAI is an agreement to liberalize the rapidly increasing flows of foreign investment around the world. Included is a broad definition of investment, including acquisitions of companies (which accounts for around 80 per cent of the category of foreign direct investment), portfolio investment (buying shares in overseas companies), intellectual property, procurement contracts by local government and natural resource concessions. The MAI would remove the rights of countries to restrict any of these types of investment. For example, Colombia does not allow foreign investment in processing hazardous and radioactive waste; Chile imposes restrictions on outward capital flows to deter speculation; Namibia does not allow unrestricted investment in natural resources, protecting its forests and inshore fisheries; many countries restrict foreign purchase of land held by smallholder farmers. These restrictions would not be allowed under the MAI. Nor would the MAI allow the types of policies that have been used by all other countries during their development phase. Every country has used government policies to help build the capacity of domestic industry, until they can compete internationally. By prohibiting such policies, the MAI risks locking the poorest countries into a low-wage, commodity-dependent future.

Perhaps most importantly from the perspective of animal welfare, the MAI would give foreign investors new powers to sue national governments in an international tribunal. The MAI does not allow discrimination against foreign investors, even unintentionally. Any policy that falls more heavily on foreign investors than on domestic companies would be open to challenge under the MAI. The test under the MAI is whether the direct or indirect effect of the policy is discriminatory. Similarly there are broad provisions under the MAI governing the expropriation of a foreign investor's assets. The term expropriation usually means the nationalization of company assets, and under customary international laws governments have to pay compensation. However, under the MAI, the term expropriation is broadened to include any law or policy 'that has the effect' of expropriation. As an example, there are a number of cases under the North American Free Trade Agreement (NAFTA), which has a similar provision to that proposed under the MAI. A US company called Ethyl Corporation took the Canadian government to NAFTA's tribunal for banning a petrol additive which research had shown to be a neuro-toxin. The case was recently settled out of court. The Canadian government lifted the ban, apologized and paid around £7 million to Ethyl. WDM and the other NGOs acting in a global coalition to stop the MAI believe that such an agreement could overturn national and local laws to protect the welfare of animals as well as the environment, workers' rights, public health and local communities.

At its heart lies the same exclusion of non-economic issues from the core of the agreement as happens in WTO, the World Bank and IMF. For example, MAI treats land as just another asset to be bought and sold. But it is not. It is crucial to livelihoods in the poorest countries, and beyond that, it is part of community, tradition and spirituality. The MAI is deeply flawed and undemocratic. We must act jointly to stop it.

The policies of economic liberalization cannot be held to account for all that is wrong in the world. However, they are deeply implicated in the disturbing trends that are making our world unsustainable, unequal and unsafe, for example:

- there is increasing poverty, particularly in rural areas, with over 1.3 billion people living in absolute poverty, more than ever before;
- there are increasing inequalities, with the ratio of income between the richest 20 per cent and the poorest 20 per cent rising from 30:1 in 1960 to 78:1 in 1994;
- more than 800 million people do not get enough to eat and one third of people in sub-Saharan Africa will not survive to the age of 40; and,
- there is a growing loss of biological diversity in the world and environmental crises are mounting.

The statistics do not start to tell the story of widespread damage and suffering. There needs to be a reassertion of the primacy of what all this economic activity is for. The aim should be to improve people's quality of life and the integrity of the world's environment. The world's most powerful governments and international agencies are facing a new challenge. It is coming from the social movements that provide the countervailing force to global liberalization. These movements operate in different arenas, involving a wide range of different methods. For example:

- The WDM and other NGOs are campaigning to reform the practices of multinational companies. WDM's recent campaign for the rights of banana workers in Costa Rica involved thousands of letters and cards sent to one of the banana multinationals, Del Monte, supported by the dumping of a tonne of banana skins outside Del Monte's offices! To its credit, Del Monte has acted quickly in response to public pressure and has agreed to recognise independent trade unions and minimize pesticide use.
- There are a myriad of initiatives building from the grassroots, from the local level. Many are exciting initiatives, starting to have a

widespread impact. They include Local Agenda 21 and community projects in the UK, as well as sustainable agriculture, community empowerment and urban regeneration in other countries.

- The MAI campaign provides a good example of a powerful campaign that has involved a wide range of trade unionists, churches, local authority councillors, MPs and MEPs (the European Parliament passed a resolution against the MAI by 437 votes to 8!), NGOs and even associations of small business and ethical businesses. The first stage of the struggle for fair investment rules has been won. The MAI was not signed at the end of April 1998 as was planned. The powerful corporate lobby is ensuring the MAI (or MAI by another name) is raised again in the WTO or elsewhere. But the tide is starting to turn. Campaigns such as the MAI, against genetically modified organisms and against cruelty to animals show the power that can be harnessed by committed individuals working together. We *can* turn back the tide of liberalization.

As the late US Supreme Court Justice, Learned Hand, said: 'It is easier to fight to retain the freedoms we possess than to regain the freedoms already lost.' We must fight to stop the MAI and other injustices. But we must also start the struggle to regain our rights that have been lost. We can and must bring this system under democratic control.

And that has to involve all of us in some way. One problem in calling these global systems to account is a deep feeling of disempowerment – that it is all too big and all too complex. But taking hold of the future for ourselves means getting involved from where we are now, in our own way. In the MAI work so far, it is not that there has been a particular campaigning tactic that has worked brilliantly but that opposition has sprung from many different sources. The MAI campaign has been a worldwide alliance and a cross-issue alliance – from French film producers to small farmers in India, from those taking direct action against companies to local authority councillors, from uncompromising radicals to those who are prepared to sit in meetings with government officials in tedious discussions over definitions. Such campaigns bring together a David that can challenge the Goliath.

But our fundamental challenge is still to capture the agenda of ideas. We have allowed the simplistic duality of market versus communism to result in a tyranny of ideas. The basic tenets of free markets, so persuasively used by Margaret Thatcher, now form the discourse that we then have to argue against. Instead of starting off with the notions of basic decency and acknowledgement of our interdependence in a shrinking world, the policy-making process is captured by questions such as whether it will harm our national 'competitiveness'. Thinking a new ideology is being made more difficult because the research agenda has been captured and research priorities oriented to funding opportunities.

As the research agenda is driven by the World Bank's agenda, or the corporate agenda, critical thought is not commercially attractive and funding is provided to serve the interests of large corporations and powerful agencies. Research was once regarded as common property, for the advancement of society, funded by governments. Now there is the corporatization of research, the conversion of educational institutions to learning factories, and the commercialization of traditional knowledge and creative endeavour.

Limits should be placed on the activities of the corporate sector. Under agreements such as the MAI, companies would be able to avoid their responsibilities to host communities and societies. The companies have globalized, but our systems of government have not. Unfortunately, far from developing new rules for ethical behaviour by companies, the MAI is going in exactly the wrong direction – it contains no binding obligations on companies whatsoever. The MAI contains only binding obligations on governments, reducing the ability of democratically elected governments to protect social and environmental goals. WDM considers the MAI is fundamentally misconceived. Please join us in the movement to '*count the value of everything and the price of only a few things*'.

21 Human Values and World Trade

Atherton Martin

I was surprised to find a contribution from Dennis Avery (Chapter 2) in this book. His views represent the antithesis of what the other contributors stand for. His chapter presents plausible-sounding misinformation that is typical of the dominant views on economics and human society and illustrates the distortion and displacement of human values that we need to challenge. I believe that truth does evolve from a direct concentration and confrontation of ideas.

Mr. Avery claims that the WTO is creating new affluence in developing countries. If this is true, it is true only because the WTO is connecting the affluent and rich in developing countries with the affluent and rich in the industrialized countries. That is all. This global alliance of wealth has exacerbated poverty in many countries. Just before the financial crisis in Mexico there was a report in the *New York Times* noting how many new millionaires there were in Mexico, and the very next week the Mexican economy collapsed. In this case as in many others, there appears to have been an inverse relationship between the number of millionaires and the health of the economy; more particularly, the extent of poverty in the country.

As if to find justification for the failure of this model to address poverty, Mr Avery argues that using biotechnology in our food systems will save the wildlands as less land will be required for food production. But this biotechnology is a continuation of the mindset that brought in the Green Revolution, which increased yields per acre while exacerbating hunger, poverty and ecological destruction. In his search for an explanation for this development dilemma, he has identified 'primitive agriculture of Africa' as the main factor contributing to food deficiency in that continent. Yet that same primitive agriculture sustained hundreds of millions of people for thousands of years without destroying the environment.

This sort of argument promotes an acceptance of values that define primitiveness as being the absence of modern technology, while accepting oppression, marginalization and poverty as the inescapable companions of progress.

This attempt to 'enslave' minds with the chains of a contorted logic that accepts the barbarity and savagery of poverty as the inevitable price of progress is, itself, barbarous, savage and wrong. Enslavement was wrong, it is wrong, it will always be wrong. It is unjust. It is primitive and, most disastrous of all, it still continues. Today, there is enslavement of the *mind* by global information systems and by persons and institutions of the kind with which Mr Avery works. I know them very well. They offer deliverance coated with logic and technology, and seek to enslave minds – and they succeed. There is the enslavement of young women and men in prostitution, and called tourism. There is the enslavement of women and children as cheap labour in export processing zones, as developing countries make vain attempts to export their way out of poverty and debt. And there is the enslavement of whole economies, whole cultures, whole countries, by the WTO, by the World Bank, by the IMF, by multinational corporations. How can Mr Avery ignore these vulgar manifestations of economics driven mad by distorted values?

What about the values that I think all of us hold dear, no matter what our culture is – the values of honesty, the values of caring, of compassion? What has happened that we cannot instantly spot when somebody is undermining the very values of humanity to perpetuate slavery, to perpetuate poverty even while they call it progress? We really must find our way back to those fundamental principles of our humanity. Because once we lose that, then we open the floodgates to anybody ready to dump the garbage of 'trade liberalization, deregulation, and even playing fields' on us, calling it progress, growth, development, affluence or whatever.

Mr Avery makes much of yield, and getting more and more out of production systems, as if there were no limits to what these systems can deliver. But getting more yield per acre is not always the best indicator of a successful farming system. Sometimes, the better farming system produces less, but is sustainable because it produces in a way that keeps the environment productive and clean; it connects agriculture with culture, so that the people's health, nutrition and cuisine are nurtured by the farming system. The lesson here is that not all growth is good. Cancer cells grow very fast, but that kind of growth is not good. We must insist on defining our goals, our path, our methods in a manner consistent with values that place people before profits, sustainability before maximization.

We must also beware of arguments that the issue is one of choice, the right to choose goods, services and governments. The issue is not just a question of choice; it is really one of decision-making, control and management of our lives. When communities have control over what kind of television images come into their homes, then we can talk about choice, because we have re-defined the terms of choice. But when the *terms* of choice are weighted heavily in favour of those who are delivering the goods, services and even governments, we do not really have a choice.

Sustainable what?

A word that is everywhere now in discussions about trade, growth and development is 'sustainability'. But there is something else that is more fundamental – *what* in our culture, *what* in our society, deserves to be sustained? Not everything. There is no perfect society. We need to take stock of what it is we are doing, what it is that is *right*, what it is that is *not quite right*, so that we do not end up sustaining the wrong things. What must be changed is probably the harder and better question to answer before we decide what we want to sustain. And part of what we have to look for is what are the underlying values and principles that are guiding the choices that we make.

We must go beyond food production and how we treat plants and animals, to questions of lifestyle, the way we live, the way we recreate, the way we work. What makes our life sustainable or not sustainable? These are hard questions to ask, but harder questions to address because our people have to contend with such notions as: more cars are better; getting from point A to point B faster is better; more office space is better, even if you have to reduce living space to accommodate it; fast food is better because it interferes less with your productivity. But do we know better?

Our task is to give concrete, practical form to those values and those structures that will sustain and ensure the survival of humanity. That sounds like a big thing. You know you can survive, but on a downward spiral that keeps getting worse. We are losing our children to drugs, to pornography, to crime, to all sorts of things that have been invented and are now part of the contemporary global notion of progress. We are told that it is not the values or the principles that are at fault. We are told that we will get what we want because tariffs have been reduced and the economy opened up to investment and trade. The logic of more being better has been institutionalized in the international trading arrangements. It is a very powerful argument that challenges a value judgment that proclaims we do not *need* more than we need.

Quite simply, we may be asking the wrong questions as we try to face this serious assault on our humanity. If we want to reinstall human values over trade, production and the organization of society, we have to ask different questions. Here are some of them:

- What is the reason for production? What is the motive behind organizing a society to produce anything? Is it to meet the *needs* of that society or to grow the economy, generate profit and expand gross domestic product? Those who want more yield obviously want to see the numbers rise. What is the purpose of agriculture – or, at least, what *was* its purpose? Its purpose – honourable and dignified – was to provide food for people so that the human element of the environment survives. There is a *big* difference between a concern for food security, ie providing food for people, and increased *yields.*
- What institutional form is the right form for us to ensure the survival of humanity? Is it private, corporate capital moving speedily from country to country, regardless of the impact on people? Or is it some form of community-based capital where people are drawing on their traditional practices – mixed farming, agroprocessing, nutrition, health, all of which were intimately linked in farming systems before, but have now become separate sectors governed by separate government ministries. They are separate now and we have to spend millions to re-connect them. But traditional farming, which was about food, had all of them wonderfully connected: education, human resource development and technology. These were an intimate and integral part of farming systems. Our grandparents had not lost sight of the fact that they were in the most dignified profession of all, which was ensuring the continuity and health of humanity by growing healthy food
- What type of politics do we need? Is it the politics of representation: you vote for a Prime Minister every four or five years? Or is it the politics of *constant* involvement of communities, institutions and people in giving direction, in monitoring, in holding accountable those who have been charged with administering the people's affairs – whether in public or private organizations?

- What technology is suitable? For some, the answer is simple – any technology that will increase yield *is* the appropriate technology, even if increasing that yield puts people out of work, destroys entire economies, pollutes ground water, introduces technologies that are totally inappropriate to the culture, the life and the survival of those communities involved. For us, what would be the value judgment on technology? Would it be a technology that can be appropriate because it is appropriated by the community? It is *theirs*. They control it, they manage it, they develop it. They massage it into shape to meet new demands. What about other technologies – how we build our homes, how we build our roads, how we care for our animals, how we care for our plants, how we trade with each other? All these are choices that are based on reminding ourselves constantly that our primary purpose is really to ensure that the life that we live is safe and satisfying and can be sustained.

If you ask the Averys of the world, 'What values drive you?', they talk about democratic society, free society, free trade, and so on, but it finally ends up – and their bottom line on this has to be – individualist, materialist gain. They are quite willing to 'let the forces of the market-place' work for them as long as it brings in more business and more money. On the other hand, a system which organizes communities to have control of their technology, control of their trade, control of information, control of the human resources development, where people can make *real* choices about trade matters or production systems – those indicators of a people-centred development do not fit into the paradigm that the Averys of the world want. (See Box 21.1 for the Dominican experience in working for human values.)

We need to focus on some simple, basic, fundamental truths which our parents seemed to have down pat. My father, who is 83 years old, clearly understood his mission in life and never missed a step. We grew up in the city, but he made sure that all of us were exposed to the rural areas. He grew plants in the house. We always had animals in the house. He ensured that we understood that everything had its place, had to be treated with respect and had to be cared for. So the dogs had to be bathed, the cats had to fed, and the chicken house had to be cleaned. When it came time to slaughter a chicken in our house – there were seven of us – *everybody* was crying and shouting. We got over it, but he had taught us a lesson. He had taught us there is a place for every living thing. Everything has a purpose, and serves its purpose, and can be handled in a way that is dignified and proper. That is a

BOX 21.1 WORKING FOR HUMAN VALUES IN DOMINICA

In Dominica, a small country of 70,000 people, we in the environmental movement use everything available to us to campaign for change. We work very closely with journalists and reporters in the media. We hold special briefing sessions for our friends in the media, so that when complex issues come up, we do not even have to talk to them. They have a range of questions that they are going to put forward and they have a range of perceptions that they are going to bring out in their reporting. Cultivating good, strong, working relationships with the media people is absolutely essential.

Secondly, we build alliances – with all groupings – from politicians to farmers, youth, women, indigenous peoples, professionals. We campaigned against BHP (Broken Hill Proprietary), the largest copper mining company in the world, which was seeking a licence to mine copper in my country. They wanted a mine that would have been one third the size of the country. We managed to prevent them even from *starting* because we forged international alliances that were able to feed us information very quickly that we put into the local media and were able to pressure our government and embarrass them into pulling out. You never know what is going to work.

A definite priority for us is working specifically with a group of people between the ages of 4 and 11. That is where 70 per cent of our work in the conservation association goes. We must build the next generation into people who do not take these things lying down, who know what injustice is and recognize it easily. We also do summer camps which help train our future scientists and researchers.

We have to be much more rigorous and thorough about the analysis we do and be able to back it up. We also use video-to-video exchange. We videotape a community meeting on a particular issue and then take that video to the next community and show it there and videotape the other community watching the video and bring that video back. Since you cannot move large numbers of people, you can move the images quickly enough through a medium that they quite understand. You get communities talking to each other so that when they actually get together on a campaign, they have established a friendship and alliances. And each community has some wonderful ideas that it can bring forth.

We certainly feel a sense of hope in the Caribbean. I find the louder the corporate world and the multinational corporations shout about the glory of trade liberalization and free markets, that is a sure signal that they are at a vulnerable point. They do not shout about the things that they do best. They shout about the things that they know *least* about because that convinces the civil servants and the policymakers that it must be correct. The truth is, they do not have answers. Their research is not rigorous, as some institutions and organizations in Europe, the US and Australia have shown. By working together, in alliances, we can assemble critical strategic groupings of people who, once they make an appearance, make the corporate world go back to their drawing boards.

The other thing is to connect that research, that information, to the communities that are most directly affected. That means translating, transposing data into formats, into methods that fishermen can understand, that farmers can understand, so they cannot be disconnected from the campaign, but become a full and constant part.

Many communities throughout the world could benefit from cooperation with highly competent researchers in biodiversity, natural resource management, learning how to take an inventory of the natural resources and finding ways to be able to use that data to monitor their management systems. These are all areas of research that translate directly into benefit for communities. We need a new alliance between researchers and communities to do that.

There have been a lot of international initiatives, not the least of which is the Global Environment Facility (GEF). This is a mixed bag because one of the implementing agencies is the World Bank. We have found that it is providing us with an interesting opportunity to hold the World Bank to account according to the GEF guidelines, which are in many cases remarkably different from the World Bank guidelines. That is allowing us to introduce a new label of accountability at the local level that we would not have had an opportunity to do before. We also need to find some ways to be able to take, as necessary, legal action against corporations whose actions harm us, to introduce a new level of local accountability into corporate operations.

value he gave us – a value that we carry into our world of work and living. Let us not allow anyone to confuse us. The key issues are not about matters of technology or yield. They are really a matter of *values*, of principles, of people. The two main messages are: nothing grows from the top down and, the magic is in the people, not the marketplace. Remember that.

Conclusion: Where Do We Go From Here?

Joyce D'Silva

By now, you may feel appalled, depressed or inspired. I hope you will be appalled at the ruthless exploitation of less economically fortunate humans, of the animals genetically mutilated in our laboratories and imprisoned in our factory farms. I hope you will be inspired by the sheer brilliance and wealth of ideas put forward by the contributors to this volume. Hopefully the inspiration will overcome any depression. Yet it is depressing to know that so many humans and animals with whom we share this planet are leading lives of pain, deprivation and suffering. Think again of the one person you know to represent the 800 million undernourished folk you do not know.

Depression is limiting. It is passive and negative. So cast it in the bin and look at what we can all do to make things better.

Philip Lymbery contrasted our functioning as consumers and as citizens. There are a wealth of ways in which we can be both more ethical consumers and campaigning citizens, seeking change.

First we must become informed. This book is a good starting place.

When we become consumers, we can learn to be label-readers. Does the small print on the egg box say 'eggs from caged hens', does the flour say 'organic'? Can we wean ourselves onto more ethically produced teas, coffees and sugars (eg Traidcraft, Oxfam)? How about toilet cleaners and washing powders that are more environmentally friendly and not tested on animals? Should we avoid the exotic mange tout flown in from Africa? Better still, why not see if there's an organic vegetable 'box' scheme in your area – or if you have a garden, grow some of your own vegetables? If you eat meat, can you demand the organic or free-range labels? Can we buy our Christmas presents from Fair Trade catalogues, or those that support an important campaign issue?

As consumers we can, in our own small way, do much. Combined with thousands, hopefully millions of others, we can become a force for change. Markets do alter in response to consumer pressure.

This book has attempted to draw together authors concerned about a whole range of interlinked issues – development, the environment, animal welfare. Your personal issue is probably included. But the message is clear: let us work together, let us make those links obvious to all. So, to buy animal-friendly toiletries and drive the most polluting car on the street does not make sense. We all need to be sensitive to *all* these vital issues.

As citizens we can support pressure groups which campaign for change, we can badger our elected representatives, we can join protest marches and rallies, we can 'spread the word' in our own milieu. We can become informed about the leviathan international agreements like GATT and the MAI, which are coming to have so much influence on the well-being of humans, animals and the environment. We can become new millennium reformers. *You* will know what is appropriate for you to do.

As Chris Fisher said from the floor of the conference, changing the rules within something like the WTO may be a long-term process, but even without changes we should not just accept the interpretations that are being placed on the existing rules. The main GATT rules existed from 1947. But it is only in the last five to ten years they have been hijacked and manipulated to reflect a new trade agenda. At meetings in spring 1998, the WTO Secretariat clearly stated that they see growth as the way forward. Yet they also recognize that growth did not necessarily bring environmental benefits or animal welfare benefits or indeed any improved equality amongst the populations of countries experiencing growth. Completely lacking is any philosophy based on redistribution, either between economies or certainly from North to South. They have not even embraced a philosophy of growth that is conditional on being at least environmentally neutral or animal welfare neutral, and poverty alleviating.

In contrast, the conference message was one of compassion – not just for animals, or just for humans, but for all sentient beings, and for the environment in which we all live out our lives. To be an 'I don't care about the human race, look what they've done to the planet, give me animals every time' sort of person both denies the real suffering of other sentient humans and gives the animal cause a bad name. Similarly, it is hard to understand those who care so much for human causes, but cannot extend their compassion to other animals – the connection just has not been made.

I remember how upset I felt when an old college friend phoned me after a piece of CIWF media publicity. She said: 'Well done – but why can't you do it for the homeless?' Of course the homeless matter, so do animals. There is no either/or. We find our energies concentrated in the area we feel most needs us – and that is fine. But we do need to be sensitive to the other issues too.

Having said that, I know so many animal campaigners who also march against the arms trade, or make tree houses to stop motorways, or help run Crisis soup rounds at Christmas. The point is clear – compassion knows *no* bounds.

Nearly all our contributors seem to sense this: Vandana Shiva, fighting for the rice farmers and for the poor leather workers, wanting the GATT rewritten to incorporate the rights of animals as well as humans; José Lutzenberger, deploring the cultural genocide that has cleared traditional peasants off their land in Brazil, but equally deploring the modern 'pig dungeons' and asking the absolutely pertinent question of agribusiness: 'Do we have a right to act as if we were the last generation?'; Mark Gold, with his call:

> *rapid progress could be made towards feeding the world healthily without exploitation of animals if only the vast resource of human intelligence and ingenuity that has been poured into wasteful projects such as conquering space or developing weapons of assured mutual destruction were instead focused passionately upon the evolution of a compassionate and rational food producing policy.*

All these people see the links, they feel the connections. Christine Townend espouses a vision of a new human–animal inter-relatedness because she feels it is better for *both*. Michael Fox sees the application of bioethics to agriculture 'so as to cause the least harm and the greatest good to the entire life community of Earth'.

This sense of community is perhaps the essence of this book. We live in an earth community of humans, animals, forests, rivers, seas and mountains. What we do can affect all. Our pollution here today can become someone else's pollution tomorrow. It is a small world in that way. Similarly, one positive act today can affect lives both near and far afield.

Compassion in World Farming Trust is concerned with 'humane' education. In its broadest sense this must mean the education of people to be sensitive to the needs of other beings, to be compassionate, to care. If this book has helped you to care more, to extend *your* compassion, then it has achieved its purpose.

Contact Addresses

Dennis T Avery
Director
Center for Global Food Issues
PO Box 202
Churchville, VA 24421, USA
Tel: +1 540 337 6354
Fax: +1 540 337 8593
Email: cgfi@rica.net

Geoffrey Cannon
Director
World Cancer Research Fund
105 Park Street
London, W1Y 3FB, UK
Tel: +44 (0)171 343 4200
Fax: + 44 (0)171 343 4201

Barry Coates
Director
World Development Movement
25 Beehive Place,
London, SW9 7QR, UK
Tel: +44 (0)171 737 6215
Fax: +44 (0)171 274 8232
Email: wdm@wdm.org.uk
Web site: http://www.wdm.org

Janice H Cox
World Animal Net
24 Barleyfields
Didcot
Oxon, OX11 0BJ, UK
Tel/Fax: +44 (0)1235 210 775
Email: worldanimalnet@ yahoo.com

Joyce D'Silva
Compassion in World Farming
Charles House
5A Charles Street, Petersfield
Hampshire, GU32 3EH, UK
Tel: +44 (0)1730 264 208 and +44 (0)1730 268 863
Fax: +44 (0)1730 260 791
Email: compassion@ciwf.co.uk

Michael Fox
The Humane Society of the United States
2100 L Street, NW
Washington, DC 20037, USA
Tel: +1 202 452 1100
Fax: +1 202 293 5109
Email: bioethics@ix.netcom.com

Maneka Gandhi
People for Animals
A–4, Maharani Bagh
New Delhi–110 065, India
Tel: + 91 11 6840402
Fax: +91 11 682144

Mark Gold
3 Ramsden Lane
Offwell
Honiton, EX14 9RZ, UK
Tel: +44 (0)1401 831763

Patrick Holden
Soil Association
Bristol House
40–56 Victoria Street
Bristol BS2 6BY, UK
Tel: +44 (0)117 929 0661
Fax: +44 (0)117 925 2504
Email: pholden@soil association.org

Tim Lang
The Centre for Food Policy
Thames Valley University
St Mary's Road
Ealing, London W5 5RP, UK
Fax: +44 (0)181 280 5125

José Lutzenberger
Rua Jacinto Gomes, 39
90040 – 270 Porto Alegre – RS, Brazil
Tel: +55 51 331 31 05
Fax: +55 51 330 35 67
Email: fundgaia@zaz.com.br

Philip Lymbery
Compassion in World Farming
Charles House
5A Charles Street, Petersfield
Hampshire, GU32 3EH, UK
Tel: +44 (0)1730 264 208 and +44 (0)1730 268 863
Fax: +44 (0)1730 260 791
Email: compassion@ciwf.co.uk

Atherton Martin
PO Box 109
Roseau Commonwealth of Dominica
West Indies
Tel: +1 767 448 8839 home
+1 767 448 4334 office
Fax: +1 767 448 8829 home
+1 767 448 4334 office
Email: exotica@cwdom.dm

Chris Mullin, MP
House of Commons
London SW1A 0AA, UK
Tel: +44 (0)191 567 2848

Hugh Raven
14 Avondale Park Gardens
London, W11 4PR, UK

Julie Sheppard
Email: sheppardj@which.co.uk

Vandana Shiva,
Research Foundation for Science, Technology and Ecology,
A–60, Hauz Khas, New Delhi – 110 016, India
Fax: +91 11 885 0795
E-mail: vandana@twn.unv.ernet.in

Peter Stevenson
Compassion in World Farming
Charles House
5A Charles Street, Petersfield
Hampshire, GU32 3EH, UK
Tel: +44 (0)1730 264 208 and +44 (0)1730 268 863
Fax: +44 (0)1730 260 791
Email: compassion@ciwf.co.uk

Geoff Tansey
Email: g.tansey@zen.co.uk

Christine Townend
Help in Suffering Animal Shelter
Maharani Farm
Durgapura, Jaipur, Rajasthan
302018, India
Fax: +91 141 548044

John Vidal
The Guardian
119 Farringdon Road
London, EC1R 3ER, UK
Tel: +44 (0)171 278 2332
Fax: +44 (0)171 837 2114
Email: VIDAL@guardian.co.uk

Mark F Watts, MEP
Transport House
Apsley St
Ashford
Kent, TN23 1LF, UK
Fax: +44 (0)1233 663 510
Email: mwatts@europarl.eu.int

Index

Page references in *italics* indicate figures or boxed text.